High-Temperature Liquid Chromatography
A User's Guide for Method Development

RSC Chromatography Monographs

Series Editors:
R.M. Smith, *Loughborough University of Technology, UK*

Advisory Panel:
J.C. Berridge, *Sandwich, UK*, G.B. Cox, *Indianapolis, USA*, I.S. Lurie, *Virginia, USA*, P.J. Schoenmakers, *Eindhoven, The Netherlands*, C.F. Simpson, *London, UK*, G.G. Wallace, *Wollongong, Australia*

Titles in this Series:
 1: Chromatographic Integration Methods
 2: Packed Column SFC
 3: Chromatographic Integration Methods, Second Edition
 4: Separation of Fullerenes by Liquid Chromatography
 5: Applications of Solid Phase Microextraction
 6: HPLC: A Practical Guide
 7: Capillary Electrochromatography
 8: Hyphenated Techniques in Speciation Analysis
 9: Cyclodextrins in Chromatography
10: Electrochemical Detection in the HPLC of Drugs and Poisons
11: Validation of Chromatography Data Systems: Meeting Business and Regulatory Requirements
12: Thin-layer Chromatography: A Modern Practical Approach
13: High-Temperature Liquid Chromatography: A User's Guide for Method Development

How to obtain future titles on publication:
A standing order plan is available for this series. A standing order will bring delivery of each new volume immediately on publication.

For further information please contact:
Book Sales Department, Royal Society of Chemistry, Thomas Graham House, Science Park, Milton Road, Cambridge, CB4 0WF, UK
Telephone: +44 (0)1223 420066, Fax: +44 (0)1223 420247,
Email: books@rsc.org
Visit our website at http://www.rsc.org/Shop/Books/

High-Temperature Liquid Chromatography
A User's Guide for Method Development

Thorsten Teutenberg
Institute of Energy and Environmental Technology, Duisburg, Germany

RSC Chromatography Monographs No. 13

ISBN: 978-1-84973-013-6
ISSN: 1757-7055

A catalogue record for this book is available from the British Library

© Thorsten Teutenberg 2010

All rights reserved

Apart from fair dealing for the purposes of research for non-commercial purposes or for private study, criticism or review, as permitted under the Copyright, Designs and Patents Act 1988 and the Copyright and Related Rights Regulations 2003, this publication may not be reproduced, stored or transmitted, in any form or by any means, without the prior permission in writing of The Royal Society of Chemistry or the copyright owner, or in the case of reproduction in accordance with the terms of licences issued by the Copyright Licensing Agency in the UK, or in accordance with the terms of the licences issued by the appropriate Reproduction Rights Organization outside the UK. Enquiries concerning reproduction outside the terms stated here should be sent to The Royal Society of Chemistry at the address printed on this page.

The RSC is not responsible for individual opinions expressed in this work.

Published by The Royal Society of Chemistry,
Thomas Graham House, Science Park, Milton Road,
Cambridge CB4 0WF, UK

Registered Charity Number 207890

For further information see our web site at www.rsc.org

Preface

This book is different from other books on liquid chromatography in a number of ways. It covers a topic which has not yet been recognised by the chromatographic community as having reached its full potential. High-temperature liquid chromatography has attracted much interest in recent years. However, there is still a widespread reluctance by industry to use temperature as an active variable for speeding up the separation process or influencing the selectivity of a separation, or to employ temperature effects for the successful hyphenation of novel detection techniques.

Temperature can be regarded an underestimated parameter, which has long been neglected as a useful tool in liquid chromatographic separations. And although it is often stated that one of the key benefits of working at high temperatures in liquid chromatography is to speed up the separation process, the use of high temperatures can also lead to the development of new hyphenation techniques, which could revolutionise chromatography as we see it today. At the time of writing, high-temperature liquid chromatographic technology has matured and better equipment is increasingly available. Appropriate heating systems are commercially available which are able to generate the fast heating rates necessary for operating a column in the temperature-programmed mode. Furthermore, columns based on silica gel which can withstand higher temperatures for an extended period are currently being introduced by various column manufacturers. Nevertheless, further technological and methodical efforts are required if the technique is to become established in more regulated environments, such as the pharmaceutical industry.

The book covers the state-of-the-art of this technology and provides not only valuable theoretical but practical information as well. Indeed, the book is not written with purely academic purposes in mind, but will also benefit the researcher at the bench who is interested in the more practical considerations of high-temperature liquid chromatography. In addition, detailed information is given on the system set-up and on the dependence of solvent properties on temperature.

RSC Chromatography Monographs No. 13
High-Temperature Liquid Chromatography: A User's Guide for Method Development
By Thorsten Teutenberg
© Thorsten Teutenberg 2010
Published by the Royal Society of Chemistry, www.rsc.org

The author has acquired a sound knowledge of this technology over recent years, and has also conducted several studies with partners from industry to validate the methods employed. Many examples from these studies are included in the book.

The book concludes with a critical outlook on the future for high-temperature liquid chromatography. An evaluation is then made of the tasks necessary to make this technology more robust.

Contents

Chapter 1	**A Brief Definition of High-Temperature Liquid Chromatography**		**1**
	1.1	What is High-Temperature Liquid Chromatography?	1
	1.2	What is a Suitable Temperature Range for High-Temperature Liquid Chromatography?	4
	1.3	Why should High Temperatures be used in Liquid Chromatography?	5
	1.4	What are the Principal Requirements of High-Temperature Liquid Chromatography?	8
Chapter 2	**System Set-up for High-Temperature Liquid Chromatography**		**15**
	2.1	The Heating System	15
	2.2	The Column	20
	2.3	The Detector	21
	2.4	The Back-Pressure Regulator	22
Chapter 3	**The Heating System**		**24**
	3.1	Preheating of the Mobile Phase	25
		3.1.1 Thermal Mismatch Broadening	25
		3.1.2 Viscous Heat Dissipation	31

RSC Chromatography Monographs No. 13
High-Temperature Liquid Chromatography: A User's Guide for Method Development
By Thorsten Teutenberg
© Thorsten Teutenberg 2010
Published by the Royal Society of Chemistry, www.rsc.org

		3.1.3	Technical Implementation of Eluent Preheating	35
		3.1.4	Experimental Verification of Eluent Preheating Efficiency	37
	3.2	Column Heating		40
		3.2.1	Air-Bath Ovens	40
		3.2.2	Water-Jacket Ovens	40
		3.2.3	Block-Heating Ovens	41
	3.3	Post-Column Cooling of the Mobile Phase		42
	3.4	Temperature Programming		43
	3.5	A Critical Comparison between Different Ovens		46
		3.5.1	Air-Bath Ovens	46
		3.5.2	Water-Jacket Ovens	48
		3.5.3	Block-Heating Ovens	48
		3.5.4	Summary	49
Chapter 4	**Mobile Phase Considerations**			**52**
	4.1	Influence of Temperature on Vapour Pressure		52
		4.1.1	Prevention of a Phase Transition using a Back-Pressure Regulator	56
		4.1.2	Prevention of a Phase Transition using a Restriction Capillary	61
	4.2	Influence of Temperature on Viscosity		64
		4.2.1	Practical Implications – The Restriction Capillary	67
		4.2.2	Practical Implications – Kinetic Aspects and Column Pressure	69
	4.3	Influence of Temperature on Static Permittivity		75
	4.4	The Water–THF System		81
	4.5	The Dortmund Data Bank		83
Chapter 5	**Suitable Stationary Phases**			**87**
	5.1	Column Bleed		89
	5.2	Investigation of Column Degradation at High Temperatures		91
	5.3	Silica-Based Stationary Phases		93
	5.4	Zirconium Dioxide Stationary Phases		97
	5.5	Titanium Dioxide Stationary Phases		101
	5.6	Polymeric Stationary Phases		104
	5.7	Other Materials		105
		5.7.1	Graphitized Carbon Column	105
		5.7.2	Thermo-Responsive Stationary Phases	107
	5.8	General Conclusions		109

Chapter 6	Method Development using Temperature as an Active Variable		114
	6.1	Special Requirements of the Heating System	114
	6.2	Special Requirements of the Column Hardware	115
	6.3	Mobile Phase Considerations	116
	6.4	Influence of Temperature on Resolution	118
		6.4.1 Influence of Temperature on Retention	119
		6.4.2 Influence of Temperature on Selectivity	125
		6.4.3 Influence of Temperature on Efficiency	128
	6.5	Method Development	131
		6.5.1 Isothermal and Isocratic Separations	131
		6.5.2 Temperature Gradient and Isocratic Separation	135
		6.5.3 Simultaneous Temperature and Solvent Gradient Separation	140
		6.5.4 Detector Optimization	143
Chapter 7	Analyte Stability		149
	7.1	Evaluation of Analyte Stability using UV Detection	150
	7.2	Influence of the Stationary Phase on Analyte Stability	152
	7.3	Definition of Critical Criteria for Analyte Stability	154
Chapter 8	Special Hyphenation Techniques		158
	8.1	Flame Ionization Detection	159
	8.2	LC-NMR	164
	8.3	Isotope Ratio Mass Spectrometry	168
	8.4	LC Taste®	174
	8.5	Drug Screening	177
Chapter 9	Critical Outlook and Future Prospects		182
	9.1	Pellicular Particles	182
	9.2	Capillary and Nano HPLC	184
	9.3	Comprehensive Two-Dimensional Liquid Chromatography	189

Appendix A Vapour Pressure Data	**193**
Appendix B Viscosity Data	**197**
Appendix C Static Permittivity Data	**201**
Subject Index	**205**

CHAPTER 1
A Brief Definition of High-Temperature Liquid Chromatography

High-temperature liquid chromatography is really a fascinating topic. Nowadays, there is renewed interest in this technique which has long been talked about.

When liquid chromatography was still young, the sky seemed the limit and the capabilities of liquid chromatography were discussed with great enthusiasm. For a practitioner who was not born during the really "hot" period of chromatography, it appeared that there were no preconceptions about the boundaries of liquid chromatography. It was in the early days of HPLC that Hesse and Engelhardt stated that temperature programming should yield the same results as solvent programming, and even concluded that this procedure would have advantages over solvent gradient elution as the solvents need not be changed.[1] Today, this statement seems to be highly innovative because temperature programming is regarded very difficult to implement in industrial applications, although a number of publications have demonstrated the feasibility of this approach.[2-50] Since this initial enthusiasm, temperature has long been neglected in liquid chromatography and has not attracted much attention. However, in separation science it is like fashion: the trends of the former decades appear again and sometimes they are presented as if they were totally new. The same seems to be true for this topic, which in fact is not that new. Nevertheless, the instrumentation has improved a lot since the early days of HPLC. Therefore, it is worthwhile revisiting this old but still very new variant of liquid chromatography.

1.1 What is High-Temperature Liquid Chromatography?

Although the question seems to be very trivial, it is not easy to give an answer. Up until now, a definition of this technique does not exist although it has emerged as the topic of many scientific meetings and symposia. I will therefore try to outline what I understand to be high-temperature liquid chromatography. Looking through the literature, a range of terms has been used – some more obvious than others:[i]

- Subcritical water chromatography[33,51–61]
- Subcritical fluid chromatography[62]
- Elevated-temperature liquid chromatography[18,63–79]
- Superheated water chromatography[80–100]
- Hot eluent liquid chromatography[101]
- (Ultra) High-temperature liquid chromatography (HT-HPLC)[102–130,13]
- Thermal aqueous liquid chromatography (TALC)[131]
- and others.[132]

It seems to be very difficult to select a temperature range and to assign this region to define high-temperature liquid chromatography, as it will be called throughout this monograph. This is immediately clear if we look at the terms given above. Subcritical fluid chromatography directly refers to the mobile phase. But at which temperature range is a liquid subcritical? Many scales are referred to water, which also plays an important role in liquid chromatography. Therefore, if water is taken as a reference, an upper temperature limit of up to 374 °C, which corresponds to the critical point of water, might be considered. Above this temperature, water becomes a supercritical fluid. But how can we distinguish high-temperature liquid chromatography from conventional or room-temperature liquid chromatography? Clearly, the subcritical region extends to the lower temperature range, which is also the domain of conventional HPLC. This contradiction can be solved when we look at the third expression, which is elevated-temperature liquid chromatography. This means that the temperature should be higher than ambient temperature. But when do we exceed the temperature limit beyond which the region of high-temperature liquid chromatography is entered? Is it 40, 50 or 60 °C? In a recent review article on high-temperature HPLC, Heinisch states that "High-temperature liquid chromatography (HTLC) is a term which refers to any separation carried out at temperatures above room temperature (typically within a range from 40 °C to 200 °C) with a mobile phase in a liquid state".[133] Personally, I think that 40 °C is too low to speak about high-temperature liquid chromatography. In my opinion, we should look at the mobile phases we are using. Since I will exclusively talk about reversed-phase liquid chromatography (RP-HPLC) in this book, relevant binary solvent systems which are used in RP-HPLC will be considered.

Again, we could take water as a reference solvent and define the normal boiling point of water as the lower temperature limit for high-temperature

---

[i]As was noted by Roger Smith during various symposia, care should be taken as to use an expression which will also be referred to by other authors. Clearly, the terms "hot eluent liquid chromatography" or "thermal aqueous liquid chromatography" are not widely used. The literature search I performed revealed that there was only one publication where these expressions were used. High-temperature liquid chromatography is by far the most popular term for this technique.

Table 1.1 Compilation of the most important physical property data of the pure substances.

Solvent	Normal boiling temperature^a [°C]	Critical pressure^a p_{cr} [bar]	Critical temperature^a T_{cr} [°C]
Water	100.0	220.5	374.2
Acetonitrile	81.9	48.3	274.9
Acetone	56.3	47.0	235.0
Ethanol	78.3	63.8	243.1
Methanol	64.6	81.0	239.5
Isopropanol	82.3	47.6	235.2
Tetrahydrofuran	66.0	51.9	267.2

^{a}Data taken from reference 129.

HPLC. The normal boiling point of a liquid indicates at what temperature the liquid will turn into a gas at atmospheric pressure. Such a phase transition has to be avoided in the whole chromatographic system because it can lead to the immediate destruction of the column and strong detector noise as will be shown later on. However, two arguments speak against this temperature. First of all, this is a very high starting point, because temperatures above 100 °C are already considered extremely high in some fields of application. The second point is that normally binary mixtures of water and an organic co-solvent are used in RP-HPLC. Typically, a separation is carried out in solvent gradient mode, if very complex samples have to be analyzed containing polar and non-polar compounds. We usually start with a high water concentration and end with a high concentration of the organic modifier. Usually, methanol and acetonitrile are the most widely used organic co-solvents in reversed-phase HPLC. Now let's have a look at the normal boiling-point temperatures of these solvents, which I have listed in Table 1.1, along with some other solvents which might also be used as modifiers.[ii]

Whereas water has the highest boiling point due to strong hydrogen bonding, the boiling points for the other solvents are much lower. Acetone already starts to boil at 56 °C. For the much more common solvents methanol and tetrahydrofuran, the normal boiling-point temperatures are 65 °C and 66 °C, respectively. This means that from the perspective of the pure components, a much lower temperature limit would be appropriate if the boiling-point temperature is taken as a reference to define the lower temperature limit of high-temperature HPLC. In this case, a lower temperature limit of about 60 °C for high-temperature liquid chromatography would be appropriate. Adjusting the

---

[ii] Although acetone is rarely used in reversed-phase HPLC due to its high UV cut-off, I will show later on that this solvent exhibits some interesting features which make it an ideal candidate for method development. Furthermore, Pat Sandra made a clear statement in his opening lecture to the HPLC symposium in Dresden 2009 that acetone would be a real alternative for acetonitrile and has some favourable chromatographic properties. Since many applications are already done using a mass spectrometer instead of a UV detector, there is nothing to worry about regarding the UV cut-off.

temperature above 60 °C then requires raising the outlet pressure of the column above the atmospheric pressure. Otherwise, a phase transition would be inevitable when a solvent gradient is run from pure water to pure acetone. *Therefore, increasing the temperature above 60 °C would mean that the domain of high-temperature liquid chromatography is entered in reversed-phase HPLC.*

Now that we have defined the lower temperature limit, we can try to define the upper temperature limit. Again, it is very helpful to have a look on the data presented in Table 1.1. Besides the normal boiling-point temperatures, I have listed the critical temperatures of these solvents. From a purely thermodynamic standpoint a liquid turns into a supercritical fluid once it is above the critical temperature. Most of the organic solvents will become a supercritical fluid around 230 °C to 240 °C, while this temperature is much higher for water. In order to define the upper temperature limit I would consider the temperature at which every solvent or solvent mixture will be in the supercritical state. Here, water clearly limits this region because it has the highest critical temperature of all solvents. Using any binary mixture comprised of water and an organic co-solvent listed in Table 1.1, the critical temperature of these mixtures is always below that of pure water. This means that the upper temperature limit is defined by the critical temperature of water. *Therefore, increasing the temperature above 374 °C would mean that the domain of supercritical fluid chromatography has been entered and we have completely exited the domain of high-temperature liquid chromatography.*

Now that the domain of high-temperature HPLC has been defined, the question "what is a suitable temperature range for high-temperature liquid chromatography?" needs to be addressed and this will be done in the next section.

1.2 What is a Suitable Temperature Range for High-Temperature Liquid Chromatography?

Although in some fields of application, temperatures as high as 60 °C will not be tolerated and are considered too high to be used, the application of temperatures as high as 370 °C with a pure water mobile phase has been reported in the literature.[88] Even if it is possible to use the complete temperature range for high-temperature HPLC, the question needs to be addressed as to what is the highest temperature which can be used in routine analysis? The requirements to make use of this technique are a stationary phase, which is stable at the highest temperature you would like to apply, and a heating system which is able to generate the desired temperature. Most conventional LC heating systems are only capable of raising the temperature to 80 °C. Although it is possible with every chromatographic system which is equipped with a column oven to enter the domain of high-temperature liquid chromatography, the region cannot be exploited further. Therefore, some instrument manufacturers have developed special heating systems which have an upper temperature limit of about 200 °C. But you also have to consider another very important aspect: the stationary phase. The stability of the stationary phase is a crucial factor, as will be shown in Chapter 5. In the last few years, column manufacturers have created silica-based reversed-phase columns with considerably improved stability. Even when HPLC was still young, the stability of silica-based columns at elevated temperatures was a real problem. It is therefore not surprising that alternative materials were already being examined at the end of the 1980s with polystyrene–divinylbenzene (PS–DVB) phases and then with the pioneering work on metal oxides, such as zirconia and titania by the groups of Carr and others.[134,135] Since then, many articles have reviewed metal oxide stationary phases designed for high-temperature HPLC.[136–147,66] Although silica-based phases have long lagged behind, they are now catching up in terms of stability at high temperatures. In some cases, they are even more stable than their metal oxide-based counterparts. From many recent studies it can be deduced that temperatures as high as 200 °C will not lead to an immediate collapse of the column, and some columns can be used over a reasonably long time without total degradation. What needs to be discussed is the question of what can be regarded as long-term stability, which will be done in Chapter 5.

Although the domain of high-temperature liquid chromatography potentially extends up to 374 °C, the useful temperature range for routine analysis that will be considered in this monograph is currently limited to approximately 200 °C. This is quite reasonable because specially designed heating systems as well as suitable stationary phases both generating and withstanding these temperatures are now commercially available.

It is undoubtedly fine to have defined what can be understood of high-temperature HPLC, but now the question has to be addressed as to why it should be beneficial to increase the temperature.

1.3 Why should High Temperatures be used in Liquid Chromatography?

In today's high speed world, it seems that all that matters is efficiency and throughput, which means a demand to reduce the time required to perform each separation. Indeed, the great majority of publications which have been written about high-temperature HPLC start by explaining the impact of temperature on the separation speed. Clearly, minimizing analysis time is one very important aspect, but it does not provide the whole story of the potential role of high-temperature HPLC. Although it might sound a little curious, temperature can be regarded as a universal parameter in liquid chromatography, even though as many authors have pointed out, it is the most underestimated parameter. Why is this? The answer is again not as straightforward as might be expected. The temperature influences almost every other parameter which can be used to optimize a separation in terms of speed *and* resolution. This means that temperature can be used to adjust the selectivity of the phase system without changing the mobile or stationary phase. Computer-optimization software has been available for a long time to assist the user in finding the optimal separation parameters, including the gradient slope as well as temperature. In our own laboratory, work is currently carried out to include even temperature programming. For the moment I will not go into further detail about this aspect, because the influence of temperature on all the important parameters related to the optimization of resolution is discussed later in Chapter 6. Instead, I would like to briefly discuss the van Deemter equation, which should be known to every chromatographer and can be written as:

$$H_u = A + \frac{B}{u} + C \cdot u \qquad (1.1)$$

Here, the Height Equivalent to a Theoretical Plate $H(u)$ (HETP) depends on three terms: the band broadening due to eddy diffusion (A-term); longitudinal diffusion (B-term); and the resistance to mass transfer between and within the mobile and stationary phases (C-term) and the mobile phase flow rate (u). Physically, it is often assumed that the A-term does not depend on temperature. However, the remaining B- and C-term are both temperature-dependent. This is because the B-term is directly proportional to the diffusion coefficient, while the C-term is inversely proportional to the diffusion coefficient (D_M), which is temperature-dependent:

$$B \propto D_M \qquad (1.2)$$

$$C \propto \frac{1}{D_M} \qquad (1.3)$$

From a purely theoretical standpoint, the goal is always to minimize band broadening and thus minimize H by adjusting the flow rate of the mobile phase to the optimum linear velocity. This is highlighted in a plot of the HETP against the linear velocity of the mobile phase in Figure 1.1.

It is obvious that at velocities higher and lower than the optimum linear velocity there is an increase of the $H(u)$-curve. However, when the temperature is increased, the profile of this curve changes. The minimum of the $H(u)$-curve is shifted to higher linear velocities. In addition, there is a much flatter increase of H at flow rates higher than the optimum. Some authors describe this as a "flattening out" of the van Deemter curve. This means that if a separation is carried out at a mobile phase flow rate which is much higher than the optimum flow rate, the loss in efficiency at higher temperatures is less pronounced than at lower temperatures. Please note that because the optimum linear velocity is shifted to higher flow rates, working at low velocities below the optimum flow rate can result in a significant loss of efficiency when the temperature is increased. This means that the flow rate should be high enough to suppress the effects of longitudinal diffusion of molecules. In industry, liquid chromatographic separations are usually not carried out at the optimum flow rate as the van Deemter curve is not recorded if a new method is created. This is practical because many separations are not critical and hence there is no need to adjust

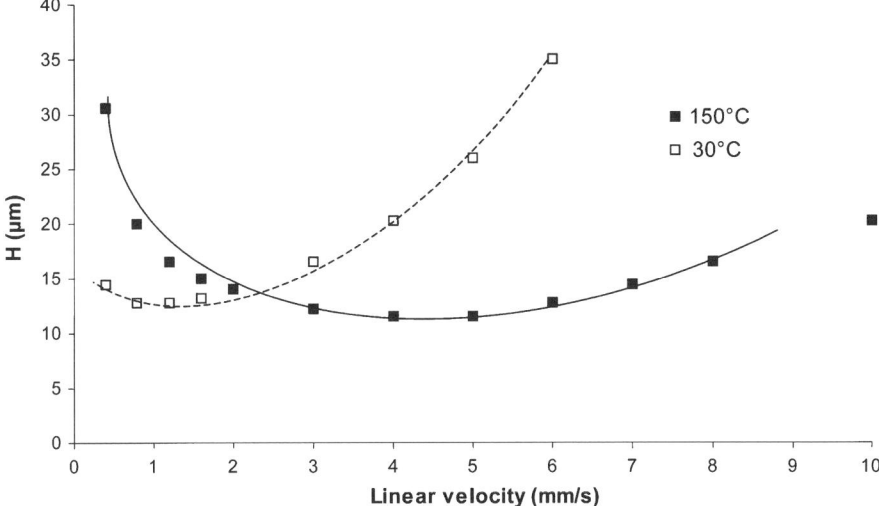

Figure 1.1 Plot of plate height (H) vs. linear velocity (u) for propylparaben (k $\sim$ 3.6 at 30 °C) on a Blaze 200 column (150 mm × 4.6 mm ID; 5 µm); mobile phase: water–acetonitrile (60 : 40); detection: DAD at 210 nm.[18] (Reproduced with kind permission from Wiley-VCH Verlag GmbH & Co. KGaA.)

the flow rate to the optimum linear velocity. *The net benefit of operating HPLC columns at higher temperatures therefore is that the operator need not worry so much about the flow rate as long as it is higher than the optimum linear velocity. However, it needs to be stressed that there is no absolute increase in the efficiency, because it is not possible to lower the minimum of the van Deemter curve.* This is often not correctly presented or misunderstood if people speak of an increase in the efficiency by increasing temperature.

Another big advantage of increasing the temperature is that the viscosity maximum of the mobile phase can be reduced. When a solvent gradient is applied, which means that the concentration of the organic co-solvent is changed from *e.g.* 0 to 100% during a chromatographic run, often a huge pressure maximum is observed. Every practitioner will have noticed that mixtures consisting of water and methanol are much more troublesome than the corresponding mixtures of water and acetonitrile at ambient temperature. However, by steadily increasing the temperature, this pressure maximum can be totally avoided, as will be shown in Chapter 4. This is because viscosity is a strong function of temperature and decreases with increasing temperature.[iii]

The effect of increased temperature is not only to increase the throughput or to decrease the viscosity of the mobile phase. Another aspect which is often overlooked is that temperature also changes the dielectric properties of the mobile phase. At ambient temperature, water is a very weak solvent if used in

---

[iii] Please note that the opposite is true if the mobile phase is a gas, as in gas chromatography. Here, the viscosity increases with increasing temperature.

reversed-phase mode. Usually, an organic solvent is necessary to elute non-polar compounds from a given stationary phase. However, increasing the temperature leads to a partial break up of the strong hydrogen-bond network between the water molecules. This in turn leads to a decrease in the water's polarity or, as it is correctly termed, in its static permittivity.[iv,v] The change of the water's polarity at higher temperatures enables us to completely eliminate the organic co-solvent. This means that separations using only water as the mobile phase can be performed on a reversed-phase column. Instead of a solvent gradient, a temperature gradient can be used to elute polar and non-polar compounds within one chromatographic run. It is therefore possible to greatly reduce the amount of toxic and costly organic solvents. Smith has made important contributions in this respect and has water termed the "ultimate green solvent".[97,100] But there are also other aspects which are equally important from a purely technical standpoint. The huge potential of these subcritical water separations enables special hyphenation techniques like the coupling of liquid chromatography with a flame ionization detector or an isotope ratio mass spectrometer to be possible. Due to the enormous potential of these new techniques, which are scarcely known even to the experienced practitioner, a complete chapter of this book has been devoted to these technologies (see Chapter 8).

At the end of this paragraph it can be summarized that the main benefit of using temperature as an active variable in liquid chromatography is not only to speed up separation or increase throughput, but to employ new hyphenation techniques which can be used to gather new information about a sample which would otherwise not be accessible by conventional liquid chromatography. These hyphenation techniques are based on the ability of water to be used as the sole eluent at high temperatures.

1.4 What are the Principal Requirements of High-Temperature Liquid Chromatography?

I would like to close this chapter with a brief description of the principal requirements for high-temperature liquid chromatography. In the following chapters, a detailed discussion will follow, but at this point it is important to gain an impression of the system set-up.

As can be deduced from the definition of high-temperature HPLC given earlier, everyone who has an HPLC system in his or her laboratory is able to perform high-temperature liquid chromatography if the system is equipped with temperature control. Most heating systems are able to control the temperature up to 80 °C, which would be sufficient to enter the regime of HTLC. This means that virtually no physical change of the existing hardware is

---

[iv] It should be noted that in the literature, the old term "dielectric constant" is often found instead of "static permittivity". From a physical point of view this is not acceptable because the static permittivity of a solvent or a mixture is always a function of pressure and temperature. Therefore, in this book, only the term static permittivity will be used instead of dielectric constant.

[v] In Chapter 4 it will be shown that this is a general phenomenon which holds also true for all binary solvent systems composed of water and an organic solvent.

necessary. Of course, special heating devices are needed if one would like to extend towards the upper temperature limit. Then, you should be aware of the need to prevent a phase transition of the mobile phase in the system. However, this can be avoided by installing a simple back-pressure regulator behind the column. Depending on the detector you would like to use, the mobile phase may need to be cooled down before it enters the detector, otherwise, irreversible damage might occur. Also, the hot mobile phase can lead to a reduction in the signal intensity for some detectors. Therefore, the use of a specially designed heating system is strongly recommended.

The next question is "which column can be used?" It is a matter of fact, that every column will degrade faster at elevated temperatures when compared to the case at room temperature. However, column manufacturers have tremendously improved the stability of silica based reversed-phase materials. Therefore, the fear of a rapid breakdown or degradation of the column is often overestimated if the temperature is kept below 80 °C. Nevertheless, it strongly depends on the specific experimental conditions which also include the pH of the mobile phase. I will give some experimental evidence which highlights the big differences in column stability of various stationary phases in Chapter 5.

For everyone who is not familiar with high-temperature HPLC, it would therefore be a good practice to just run some methods at elevated temperatures. As I have outlined above, virtually no changes have to be made if you would like to increase the temperature up to 60 °C. Therefore, before going ahead with the study of this book, go into the laboratory and run a method at 60 °C. In this case you should not observe any problems due to boiling of the mobile phase when using methanol or acetonitrile as the organic co-solvent. Nevertheless, study the advice given by the column manufacturer for the column you are using. There are some materials which are extremely prone to degradation and 60 °C might already lead to a partial breakdown of the stationary phase. However, most modern C18 silica-based stationary phases are stable at this temperature. When you have carried out a few experiments, you can then continue to the other chapters. You will see that it is then much easier to follow the discussion outlined in this book because some problems and questions only arise when you return from the laboratory, rather than just using this book for mere theoretical study. If you are already experienced with working at high eluent temperatures and are familiar with the hardware you are using, then it should be no problem to follow up the discussion.

So, just to sum it up: if you apply temperatures no higher than 60 °C, then you can use the conventional system set-up with virtually no modification. In case you wish to explore the real high-temperature region, a back-pressure control, a suitable heating device and a stable stationary phase are needed. In the next chapter, I will go into more detail and explain the system set-up for high-temperature HPLC.

References

1. G. Hesse and H. Engelhardt, *J. Chromatogr.*, 1966, **21**, 228.
2. L. R. Snyder, *J. Chromatogr. Sci.*, 1970, **8**, 692.

3. E. J. Kikta, A. E. Stange and S. Lam, *J. Chromatogr.*, 1977, **138**, 321.
4. J. Bowermaster and H. McNair, *J. Chromatogr.*, 1983, **279**, 431.
5. J. Bowermaster and H. M. McNair, *J. Chromatogr. Sci.*, 1984, **22**, 165.
6. W. R. Biggs and J. C. Fetzer, *J. Chromatogr.*, 1986, **351**, 313.
7. H. McNair and J. Bowermaster, *J. High Resolut. Chromatogr. Chromatogr. Commun.*, 1987, **10**, 27.
8. S. U. Sheikh and J. C. Touchstone, *J. Chromatogr.*, 1988, **455**, 327.
9. K. Jinno and M. Yamagami, *Chromatographia*, 1989, **27**, 417.
10. T. Okada, *Anal. Chem.*, 1991, **63**, 1043.
11. K. Ryan, N. M. Djordjevic and F. Erni, *J. Liq. Chromatogr. Relat. Technol.*, 1996, **19**, 2089.
12. M. H. Chen and C. Horvath, *J. Chromatogr. A*, 1997, **788**, 51.
13. N. M. Djordjevic, P. W. J. Fowler and F. Houdiere, *J. Microcolumn Sep.*, 1999, **11**, 403.
14. R. Trones, T. Andersen, T. Greibrokk and D. R. Hegna, *J. Chromatogr. A*, 2000, **874**, 65.
15. R. Trones, T. Andersen, D. R. Hegna and T. Greibrokk, *J. Chromatogr., A*, 2000, **902**, 421.
16. T. Andersen, P. Molander, R. Trones, D. R. Hegna and T. Greibrokk, *J. Chromatogr., A*, 2001, **918**, 221.
17. T. Greibokk and T. Andersen, *J. Sep. Sci.*, 2001, **24**, 899.
18. G. Vanhoenacker and P. Sandra, *J. Sep. Sci.*, 2006, **29**, 1822.
19. Y. Hirata and K. Jinno, *J. High Resolut. Chromatogr. Chromatogr. Commun.*, 1983, **6**, 196.
20. A. F. Bergold and P. W. Carr, *Anal. Chem.*, 1989, **61**, 1117.
21. J. S. Yoo, J. T. Watson and V. L. McGuffin, *J. Microcolumn Sep.*, 1992, **4**, 349.
22. K. Van Lenning, J. L. Garrido, J. Aristegui and M. Zapata, *Chromatographia*, 1995, **41**, 539.
23. N. M. Djordjevic, F. Houdiere, P. F. Fowler and F. Natt, *Anal. Chem.*, 1998, **70**, 1921.
24. P. Molander, K. Haugland, D. R. Hegna, E. Ommundsen, E. Lundanes and T. Greibrokk, *J. Chromatogr., A*, 1999, **864**, 103.
25. P. Molander, S. J. Thommesen, I. A. Bruheim, R. Trones, T. Greibrokk, E. Lundanes and T. E. Gundersen, *J. High Resolut. Chromatogr.*, 1999, **22**, 490.
26. I. Bruheim, P. Molander, E. Lundanes, T. Greibrokk and E. Ommundsen, *J. High Resolut. Chromatogr.*, 2000, **23**, 525.
27. P. Molander, K. Haugland, E. Lundanes, S. Thorud, Y. Thomassen and T. Greibrokk, *J. Chromatogr., A*, 2000, **892**, 67.
28. P. Molander, A. Holm, E. Lundanes, T. Greibrokk and E. Ommundsen, *J. High Resolut. Chromatogr.*, 2000, **23**, 653.
29. I. Bruheim, P. Molander, M. Theodorsen, E. Ommundsen, E. Lundanes and T. Greibrokk, *Chromatographia*, 2001, **53**, S266.
30. P. Molander, T. Greibrokk, A. Iveland and E. Ommundsen, *J. Sep. Sci.*, 2001, **24**, 136.

31. P. Molander, A. Thomassen, L. Kristoffersen, T. Greibrokk and E. Lundanes, *J. Chromatogr. B*, 2001, **766**, 77.
32. T. Andersen, I. L. Skuland, A. Holm, R. Trones and T. Greibrokk, *J. Chromatogr., A*, 2004, **1029**, 49.
33. T. Teutenberg, H. J. Goetze, J. Tuerk, J. Ploeger, T. K. Kiffmeyer, K. G. Schmidt, W. G. Kohorst, T. Rohe, H. D. Jansen and H. Weber, *J. Chromatogr., A*, 2006, **1114**, 89.
34. J. P. Godin, G. Hopfgartner and L. Fay, *Anal. Chem.*, 2008, **80**, 7144.
35. H. C. Lee and T. Y. Chang, *Polymer*, 1996, **37**, 5747.
36. F. Houdiere, P. W. J. Fowler and N. M. Djordjevic, *Anal. Chem.*, 1997, **69**, 2589.
37. W. Lee, H. C. Lee, T. Y. Chang and S. B. Kim, *Macromolecules*, 1998, **31**, 344.
38. T. Y. Chang, H. C. Lee, W. Lee, S. Park and C. H. Ko, *Macromol. Chem. Phys.*, 1999, **200**, 2188.
39. W. Lee, H. C. Lee, T. Park, T. Chang and J. Y. Chang, *Polymer*, 1999, **40**, 7227.
40. W. Lee, H. C. Lee, T. Park, T. Chang and K. H. Chae, *Macromol. Chem. Phys.*, 2000, **201**, 320.
41. W. Lee, D. Cho, B. O. Chun, T. Chang and M. Ree, *J. Chromatogr., A*, 2001, **910**, 51.
42. N. Rosales-Conrado, M. E. Leon-Gonzalez, L. V. Perez-Arribas and L. M. Polo-Diez, *Anal. Chim. Acta*, 2002, **470**, 147.
43. A. Holm, P. Molander, E. Lundanes and T. Greibrokk, *J. Sep. Sci.*, 2003, **26**, 1147.
44. P. Molander, R. Olsen, E. Lundanes and T. Greibrokk, *Analyst*, 2003, **128**, 1341.
45. I. L. Skuland, T. Andersen, R. Trones, R. B. Eriksen and T. Greibrokk, *J. Chromatogr., A*, 2003, **1011**, 31.
46. J. Ryu, S. Park and T. Y. Chang, *J. Chromatogr., A*, 2005, **1075**, 145.
47. S. Teramachi, H. Matsumoto and T. Kawai, *J. Chromatogr., A*, 2005, **1100**, 40.
48. S. Park and T. Chang, *Macromolecules*, 2006, **39**, 3466.
49. S. Park, C. Ko, H. Choi, K. Kwon and T. Chang, *J. Chromatogr., A*, 2006, **1123**, 22.
50. C. Y. Shih, Y. Chen, J. Xie, Q. He and Y. C. Tai, *J. Chromatogr., A*, 2006, **1111**, 272.
51. D. J. Miller and S. B. Hawthorne, *Anal. Chem.*, 1997, **69**, 623.
52. Y. Yang, A. D. Jones and C. D. Eaton, *Anal. Chem.*, 1999, **71**, 3808.
53. Y. Yang, A. D. Jones, J. A. Mathis and M. A. Francis, *J. Chromatogr., A*, 2002, **942**, 231.
54. Y. Yang, L. J. Lamm, P. He and T. Kondo, *J. Chromatogr. Sci.*, 2002, **40**, 107.
55. P. He and Y. Yang, *J. Chromatogr., A*, 2003, **989**, 55.
56. T. Kondo and Y. Yang, *Anal. Chim. Acta*, 2003, **494**, 157.
57. L. J. Lamm and Y. Yang, *Anal. Chem.*, 2003, **75**, 2237.

58. S. Yisong, J. Jen-Fon and Z. Weibing, *Chin. J. Chromatogr.*, 2005, **23**, 238.
59. M. O. Fogwill and K. B. Thurbide, *J. Chromatogr.*, 2007, **1139**, 199.
60. Y. Yang, *J. Sep. Sci.*, 2007, **30**, 1131.
61. M. O. Fogwill and K. B. Thurbide, *J. Chromatogr., A*, 2008, **1200**, 49.
62. C. West and E. Lesellier, *J. Chromatogr., A*, 2005, **1099**, 175.
63. S. Abbott, P. Achener, R. Simpson and F. Klink, *J. Chromatogr.*, 1981, **218**, 123.
64. H. Chen and C. Horvath, *Anal. Meth. Instrum.*, 1993, **1**, 213.
65. X. C. Le, M. Ma and N. A. Wong, *Anal. Chem.*, 1996, **68**, 4501.
66. J. W. Li, Y. Hu and P. W. Carr, *Anal. Chem.*, 1997, **69**, 3884.
67. G. C. Sheng, Y. F. Shen and M. L. Lee, *J. Microcolumn Sep.*, 1997, **9**, 63.
68. X. C. Le, X.-F. Li, V. Lai, M. Ma, S. Yalcin and J. Feldmann, *Spectrochim. Acta,Part B*, 1998, **53**, 899.
69. U. D. Neue and J. R. Mazzeo, *J. Sep. Sci.*, 2001, **24**, 921.
70. J. D. Thompson, J. S. Brown and P. W. Carr, *Anal. Chem.*, 2001, **73**, 3340.
71. A. Jones and Y. Yang, *Anal. Chim. Acta*, 2003, **485**, 51.
72. B. A. Bidlingmeyer and J. Henderson, *J. Chromatogr., A*, 2004, **1060**, 187.
73. J. W. Coym and J. G. Dorsey, *J. Chromatogr., A*, 2004, **1035**, 23.
74. C. R. Zhu, D. M. Goodall and S. A. C. Wren, *LCGC Eur.*, 2004, **17**, 530.
75. Y. Ueki, T. Umemura, Y. Iwashita, T. Odake, H. Haraguchi and K. Tsunoda, *J. Chromatogr., A*, 2006, **1106**, 106.
76. Y. Xiang, Y. Liu and M. L. Lee, *J. Chromatogr., A*, 2006, **1104**, 198.
77. Z. Hao, C. Y. Lu, B. Xiao, N. Weng, B. Parker, M. Knapp and C. T. Ho, *J. Chromatogr., A*, 2007, **1147**, 165.
78. J. A. Lippert, T. M. Johnson, J. B. Lloyd, J. P. Smith, B. T. Johnson, J. Furlow, A. Proctor and S. J. Marin, *J. Sep. Sci.*, 2007, **30**, 1141.
79. P. Sandra and G. Vanhoenacker, *J. Sep. Sci.*, 2007, **30**, 241.
80. R. M. Smith and R. J. Burgess, *Anal. Commun.*, 1996, **33**, 327.
81. R. M. Smith and R. J. Burgess, *J. Chromatogr., A*, 1997, **785**, 49.
82. B. A. Ingelse, H.-G. Janssen and C. A. Cramers, *J. High Resolut. Chromatogr.*, 1998, **21**, 613.
83. O. Chienthavorn and R. M. Smith, *Chromatographia*, 1999, **50**, 485.
84. R. M. Smith, R. J. Burgess, O. Chienthavorn and J. R. Bone, *LCGC Int.*, 1999, **17**, 938.
85. I. D. Wilson, *Chromatographia*, 2000, **52**, 28.
86. S. M. Fields, C. Q. Ye, D. D. Zhang, B. R. Branch, X. J. Zhang and N. Okafo, *J. Chromatogr., A*, 2001, **913**, 197.
87. T. Teutenberg, O. Lerch, H. J. Götze and P. Zinn, *Anal. Chem.*, 2001, **73**, 3896.
88. T. S. Kephart and P. K. Dasgupta, *Talanta*, 2002, **56**, 977.
89. R. Tajuddin and R. M. Smith, *Analyst*, 2002, **127**, 883.
90. R. Nakajima, T. Yarita and M. Shibukawa, *Bunseki Kagaku*, 2003, **52**, 305.
91. T. Yarita, R. Nakajima and M. Shibukawa, *Anal. Sci.*, 2003, **19**, 269.

92. O. Chienthavorn, R. M. Smith, S. Saha, I. D. Wilson, B. Wright, S. D. Taylor and E. M. Lenz, *J. Pharm. Biomed. Anal.*, 2004, **36**, 477.
93. J. W. Coym and J. G. Dorsey, *Anal. Lett.*, 2004, **37**, 1013.
94. O. Chienthavorn, R. M. Smith, I. D. Wilson, B. Wright and E. M. Lenz, *Phytochem. Anal.*, 2005, **16**, 217.
95. R. Tajuddin and R. M. Smith, *J. Chromatogr., A*, 2005, **1084**, 194.
96. T. Yarita, R. Nakajima, K. Shimada, S. Kinugasa and M. Shibukawa, *Anal. Sci.*, 2005, **21**, 1001.
97. R. M. Smith, *Anal. Bioanal. Chem.*, 2006, **385**, 419.
98. P. Dugo, K. Buonasera, M. L. Crupi, F. Cacciola, G. Dugo and L. Mondello, *J. Sep. Sci.*, 2007, **30**, 1125.
99. L. Al-Khateeb and R. M. Smith, *J. Chromatogr., A*, 2008, **1201**, 61.
100. R. M. Smith, *J. Chromatogr., A*, 2008, **1184**, 441.
101. T. S. Kephart and P. K. Dasgupta, *Anal. Chim. Acta*, 2000, **414**, 71.
102. T. Teutenberg, S. Wiese, P. Wagner and J. Gmehling, *J. Chromatogr., A*, 2009, **1216**, 8480.
103. S. Yamaki, T. Isobe, T. Okuyama and T. Shinoda, *J. Chromatogr., A*, 1996, **728**, 189.
104. N. M. Djordjevic, F. Houdiere and P. Fowler, *Biomed. Chromatogr.*, 1998, **12**, 153.
105. R. Trones, T. Andersen, I. Hunnes and T. Greibrokk, *J. Chromatogr., A*, 1998, **814**, 55.
106. P. Molander, R. Trones, K. Haugland and T. Greibrokk, *Analyst*, 1999, **124**, 1137.
107. R. Trones, T. Andersen and T. Greibrokk, *J. High Resolut. Chromatogr.*, 1999, **22**, 283.
108. D. Louden, A. Handley, S. Taylor, I. Sinclair, E. Lenz and I. D. Wilson, *Analyst*, 2001, **126**, 1625.
109. M. M. Sanagi, H. H. See, W. A. Ibrahim and A. A. Naim, *J. Chromatogr., A*, 2004, **1059**, 95.
110. M. M. Sanagi, S. H. Heng, W. A. W. Ibrahim and A. A. Naim, *Malaysian J. Chem.*, 2004, **6**, 55.
111. M. Albert, G. Cretier, D. Guillarme, S. Heinisch and J. L. Rocca, *J. Sep. Sci.*, 2005, **28**, 1803.
112. D. Guillarme and S. Heinisch, *Sep. Purificat. Technol.*, 2005, **34**, 181.
113. G. Vanhoenacker, P. Sandra and J. Chromatogr, *J. Chromatogr., A*, 2005, **1082**, 193.
114. X. Q. Yang, L. J. Ma and P. W. Carr, *J. Chromatogr., A*, 2005, **1079**, 213.
115. L. Pereira, *LCGC North Am.-Appl. Noteb.*, 2006, **June 2006**, 75.
116. L. A. Riddle and G. Guiochon, *J. Chromatogr., A*, 2006, **1137**, 173.
117. S. Shen, H. Lee, J. McCaffrey, N. Yee, C. Senanayake and N. Grinberg, *J. Liq. Chromatogr. Relat. Technol.*, 2006, **29**, 2823.
118. D. R. Stoll, J. D. Cohen and P. W. Carr, *J. Chromatogr., A*, 2006, **1122**, 123.
119. F. Gritti and G. Guiochon, *J. Chromatogr., A*, 2007, **1169**, 125.

120. C. V. McNeff, B. Yan, D. R. Stoll and R. A. Henry, *J. Sep. Sci.*, 2007, **30**, 1672.
121. D. T. T. Nguyen, D. Guillarme, S. Heinisch, M. P. Barrioulet, J. L. Rocca, S. Rudaz and J. L. Veuthey, *J. Chromatogr., A*, 2007, **1167**, 76.
122. L. Pereira, S. Aspey and H. Ritchie, *J. Sep. Sci.*, 2007, **30**, 1115.
123. R. Plumb, J. R. Mazzeo, E. S. Grumbach, P. Rainville, M. Jones, T. Wheat, U. D. Neue, B. Smith and K. A. Johnson, *J. Sep. Sci.*, 2007, **30**, 1158.
124. S. Giegold, *Application of High Temperature-High Performance Liquid Chromatography (HT-HPLC): About the Influence of Temperature on the Analysis Time, Selectivity and different Detection Systems,* PhD thesis, University of Siegen, Germany, 2008.
125. S. Giegold, T. Teutenberg, J. Tuerk, T. Kiffmeyer and B. Wenclawiak, *J. Sep. Sci.*, 2008, **31**, 3497.
126. H. G. Gika, G. Theodoridis, J. Extance, A. M. Edge and I. D. Wilson, *J. Chromatogr., B: Biomed. Appl.*, 2008, **871**, 279.
127. S. Heinisch, G. Desmet, D. Clicq and J. L. Rocca, *J. Chromatogr., A*, 2008, **1203**, 124.
128. T. Teutenberg, *Anal. Chim. Acta*, 2009, **643**, 1.
129. T. Teutenberg, P. Wagner and J. Gmehling, *J. Chromatogr., A*, 2009, **1216**, 6471.
130. T. Teutenberg, S. Wiese, P. Wagner and J. Gmehling, *J. Chromatogr., A*, 2009, **1216**, 8470.
131. C. L. Guillemin, J. L. Millet and J. Dubois, *J. High Resolut. Chromatogr. Chromatogr. Commun.*, 1981, **4**, 280.
132. T. Kondo, Y. Yang and L. Lamm, *Anal. Chim. Acta*, 2002, **460**, 185.
133. S. Heinisch and J. L. Rocca, *J. Chromatogr., A*, 2009, **1216**, 642.
134. M. P. Rigney, T. P. Weber and P. W. Carr, *J. Chromatogr.*, 1989, **484**, 273.
135. J. Winkler and S. Marme, *J. Chromatogr., A*, 2000, **888**, 51.
136. J. Nawrocki, M. P. Rigney, A. McCormick and P. W. Carr, *J. Chromatogr., A*, 1993, **657**, 229.
137. J. Nawrocki, C. Dunlap, A. McCormick and P. W. Carr, *J. Chromatogr., A*, 2004, **1028**, 1.
138. J. Nawrocki, C. Dunlap, J. Li, J. Zhao, C. V. McNeff, A. McCormick and P. W. Carr, *J. Chromatogr., A*, 2004, **1028**, 31.
139. T. P. Weber, P. T. Jackson and P. W. Carr, *Anal. Chem.*, 1995, **67**, 3042.
140. J. Li and P. W. Carr, *Anal. Chem.*, 1996, **68**, 2857.
141. J. Li and P. W. Carr, *Anal. Chem.*, 1997, **69**, 2202.
142. J. Li and P. W. Carr, *Anal. Chem.*, 1997, **69**, 2193.
143. J. Li and P. W. Carr, *Anal. Chem.*, 1997, **69**, 837.
144. B. Yan, J. Zhao, J. S. Brown, J. Blackwell and P. W. Carr, *Anal. Chem.*, 2000, **72**, 1253.
145. Y. Mao and P. W. Carr, *Anal. Chem.*, 2000, **72**, 110.
146. Y. Xiang, B. Yan, B. Yue, C. V. McNeff, P. W. Carr and M. L. Lee, *J. Chromatogr., A*, 2003, **983**, 83.
147. D. R. Stoll and P. W. Carr, *J. Am. Chem. Soc.*, 2005, **127**, 5034.

CHAPTER 2
System Set-up for High-Temperature Liquid Chromatography

In Chapter 1 I defined what can be understood as high-temperature liquid chromatography. Furthermore, it was shown that it is possible to use any conventional HPLC system which is equipped with a temperature control to enter the domain of high-temperature HPLC. Nevertheless, using a heating system which is well adapted to higher eluent temperatures guarantees that there is no loss in efficiency due to an improper preheating of the mobile phase and that the full temperature range can be exploited. In this chapter, I will focus on the most important hardware components. The reader should be aware that I will present only the general requirements for high-temperature HPLC which were introduced in Chapter 1. This is in order to avoid initially overloading the practitioner, who is not that familiar with this technology, with too much detail.

2.1 The Heating System

In principle, every HPLC system should be equipped with a column oven. Exposing a column to ambient air can cause unwanted retention time shifts. Even if the laboratory is air-conditioned, there are often small changes in the temperature profile. It is not unusual that the difference in temperature between the night and the day can be as high as 10 °C. Such a pronounced temperature difference can lead to methods with a poor reproducibility. But even a much lower temperature difference can lead to a significant change in resolution. The example which is depicted in Figure 2.1 highlights that a temperature difference of about 6 °C leads to a marked reduction in the resolution of critical peak pairs.[i]

---

[i] Please note that with the advent of small particle packed columns and fast methods, a precise temperature control is of utmost importance, because the peak width can be easily decreased to one second and the smallest change in retention can have a detrimental effect on the overall separation efficiency.

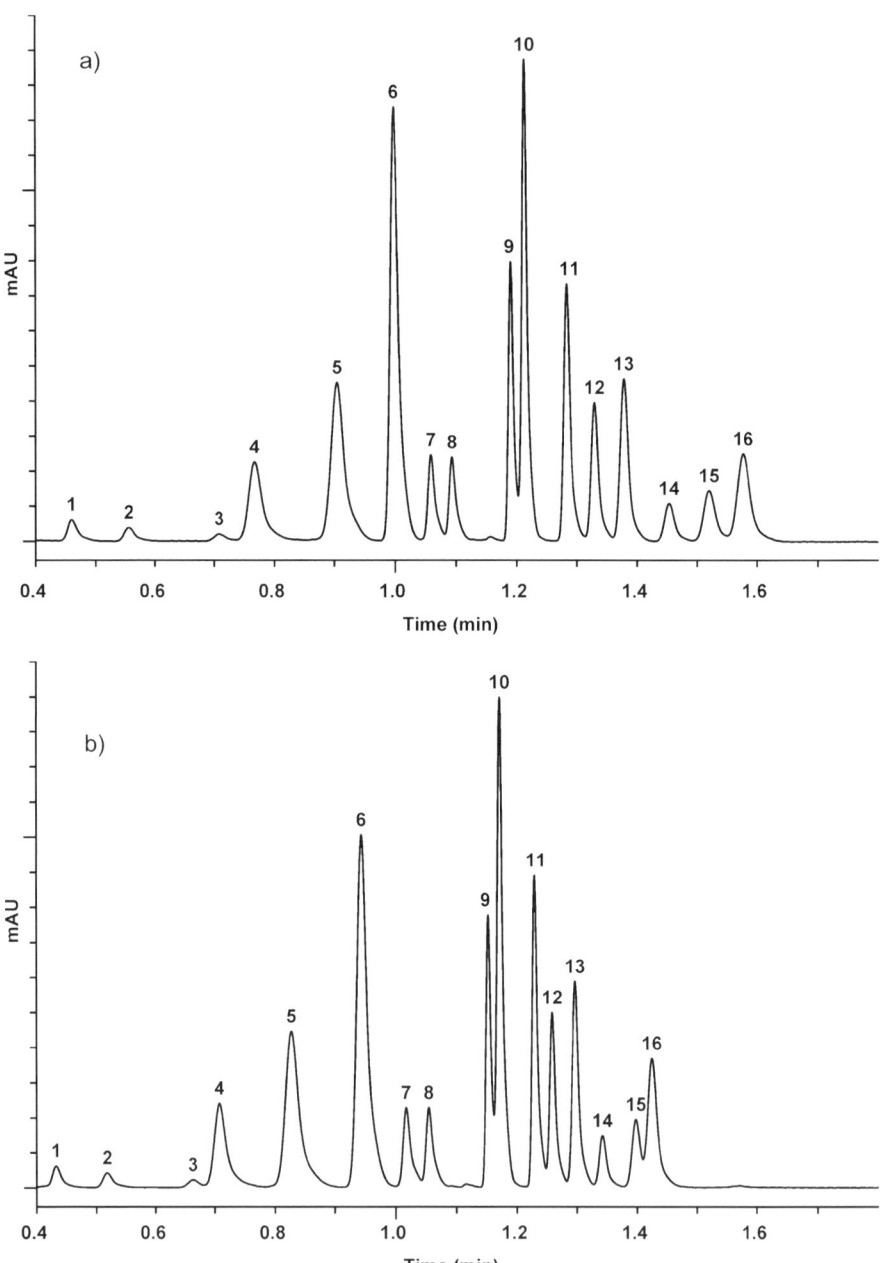

Figure 2.1 Separation of sixteen PAH on a YMC column (50 mm × 2.1 mm ID; 3 μm) with Agilent Infinity 1290 at (a) 24.3 °C and (b) 30 °C. Chromatographic conditions: solvent A: water; solvent B: acetonitrile; solvent gradient: 50 to 58%B in 0.74 min, then 58 to 100%B from 0.74 to 1.50 min; flow rate: 1.4 ml min^{-1}; injection vol.: 0.5 μl; maximum pressure during solvent gradient: 310 bar; detection: UV at 254 nm.

The separation of 16 polycyclic aromatic hydrocarbons (PAH) was achieved in less than two minutes.[1] The resolution of the critical peak pair 9/10 was 1.4 at 24.3 °C, which was found to be the optimum temperature. An increase in the temperature to 30 °C led to a critical resolution of 0.9 for peak pair 15/16. Although this is not a high-temperature separation, the example clarifies that precise temperature control is of utmost importance, even if the separation is carried out at ambient temperature. Please note that UV detection was used. If you use fluorescence detection, as is mostly employed for this kind of application to lower the limit of detection, and have defined intervals in which certain peaks should be detected this might lead to a complete failure of the method. The reason is that many fluorescence detectors only allow the adjustment of one excitation and emission wavelength at a time. In order to increase the sensitivity for all analytes, the excitation and emission wavelengths are changed during the chromatographic run. The switching of wavelengths is made at a fixed retention time based on the optimized method. A higher or lower temperature can cause a shift in the retention of analytes so that they elute at shorter or longer times. This can lead to a complete failure of the method as an accurate quantification or even detection of some compounds is not possible. The same holds true for mass spectrometric detection when you work in multiple reaction monitoring mode and have defined periods in which certain analytes should be detected. During the day you might observe that peaks will elute at a shorter retention time than during the night when the temperature is lower. When a routine assay is performed it might be found that certain compounds will only be detected during the day, but never during the night. *I strictly advise the practitioner to use a column oven, even if a separation is carried out at ambient temperature. It should be noted that ambient temperature is not defined and may vary significantly between night and day or between laboratories.*

Nevertheless, even with a column oven the problem of shifting retention times may not be completely ruled out. In the last two decades the heating system played a minor role and was often used only carelessly. Neither the instrument manufacturers nor the practitioners cared much about this issue. If you look to your HPLC system, then how important do you consider temperature control? I think that most systems are equipped with a simple air-bath oven where the column is just put inside. If you use an air-bath oven, is the air even circulated to achieve a better heat transfer? Have you ever thought about mobile phase preheating? Have you ever thought about placing a preheating capillary before your column? Have you ever thought that this preheating capillary should be of a well defined length and internal diameter to guarantee that the mobile phase will be brought to the same temperature as the column? I guess that most practitioners will answer all these questions with "no". Indeed, if you only perform the analyses at room temperature, these questions might not be that important. However, if you know that there are temperature shifts in your laboratory and you carefully re-examine your data, you might well come across some of the problems I have mentioned above. This means that a heating system should allow for preheating of the mobile phase.

Nowadays, there is basically no controversy about mobile phase preheating, especially if we talk about high-temperature HPLC. Of course, if we look at the details the problem is indeed vastly more complicated. However, for the moment it is enough if you keep in mind that the temperature of the mobile phase entering the column should exactly match the temperature of the stationary phase, although there are some exceptions from this rule. I will resume this discussion in Chapter 3.1.2, where the problem of frictional heating at very high pressures is discussed.

In order to highlight what happens if the mobile phase is not preheated while the column is operated at a very high temperature, Figure 2.2 should serve as a good illustration. This experiment was performed to visualize the effect of peak distortion at high eluent temperatures.[1] A test mixture containing eight polycyclic aromatic hydrocarbons was eluted on a Waters XBridge BEH-C_{18} stationary phase using an isocratic mobile phase of water and methanol. The upper chromatogram in Figure 2.2 resulted if the mobile phase was adequately preheated to the temperature of the stationary phase. In contrast to this, the lower chromatogram in Figure 2.2 was obtained if the mobile phase was not preheated before it entered the column. I think that this example speaks for itself and undoubtedly underscores that mobile phase preheating is a must, especially when columns with a diameter of 2 mm or greater are used. There might be people who claim that eluent preheating is not important. In such a case, the chromatographic conditions should be carefully analyzed. It might indeed be possible that these effects are not that pronounced when capillary columns are used at very low flow rates. However, the best way to really evaluate which statement is wrong or right, is to perform the experiment as described above.

The next problem concerns the control of the eluent temperature when the hot mobile phase leaves the column, because the detector signal might be influenced. The fluorescence detector is a very good example because the fluorescence signal can be very dependent on temperature. Often, a loss in signal intensity will be observed due to quenching effects when the temperature of the mobile phase entering the detector is increased.[2] Therefore, a heating system should also allow for a cool-down of the mobile phase after it leaves the column.

Now, look at your system and think of how to overcome this problem if you would like to operate the column at the maximum temperature of your column oven. In the scientific literature improvised set-ups are described. Many authors have used an ice bath in which the capillary from the column outlet to the detector was placed. If you will only make some preliminary experiments for proof of concept, then an ice bath will suffice. However, this is not a set-up I would recommend to use in a routine environment or a professional laboratory. The ice has to be replaced very often and it is not guaranteed that the mobile phase is cooled down to a constant temperature. Therefore, temperature fluctuations will be observed. This means that the heating system should be equipped with a device to cool down the eluent to a constant temperature.

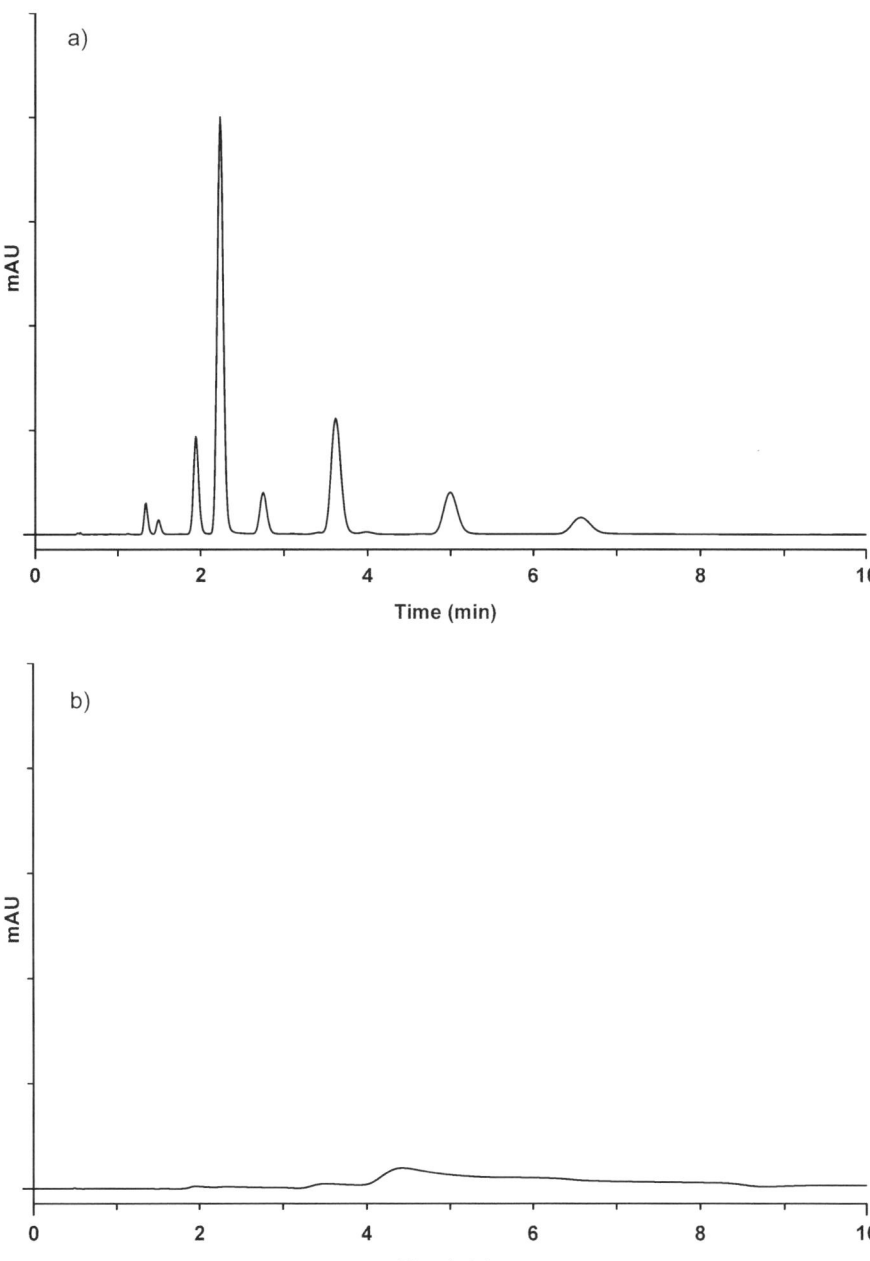

Figure 2.2 Effect of eluent preheating at very high temperature. Separation of eight PAH on a Waters XBridge BEH-C_{18} column (75 mm × 4.6 mm ID; 2.5 μm) with Shimadzu LC-10 and SIM HT-HPLC 200 column oven. Chromatographic conditions: mobile phase: water–methanol, (50 : 50 v/v); flow rate: 1.5 ml min^{-1}; injection vol.: 3 μl; column temp.: 150 °C; detection: UV at 254 nm. (See text for further details.)

Although it was stated in the first chapter that – at least theoretically – it is possible to use every conventional HPLC system with temperature control to perform high-temperature HPLC, it is obvious that some modifications are necessary to get reproducible and high quality results. *It can be summarized at this point that a heating system should allow for the preheating of the mobile phase before it enters the column. Ideally, the temperature of the incoming mobile phase should be the same as the temperature of the column. Also, the heating system should be equipped with an integrated device to cool down the eluent to a constant temperature when it leaves the column. This guarantees that the detector signal will not be influenced by fluctuations in the eluent's temperature.*

I will enlarge upon this discussion on the heating system in the third chapter, but for the moment you need only keep in mind these facts as a rule of thumb.

2.2 The Column

The column can be regarded as the Achilles heel of HPLC and is the part which has to be replaced most often in the whole system. However, many users still think that the column should last forever. Indeed, there is a lot of on-going controversy about the interval before a column has to be replaced. As was already outlined in the first chapter, one aspect of high-temperature HPLC is to increase the sample throughput by accelerating the separation process. Therefore, if a column has to be replaced after 20 instead of 40 working days, this does not necessarily mean that the cost per sample is higher. Let's assume that it is possible to decrease the overall analysis time by a factor of four just by increasing the temperature. A decrease in the analysis time by a factor of four and the decrease in the column lifetime by a factor of two means that you have increased the sample throughput by a factor of four, and that the number of samples which can be analyzed on the same column has been doubled before the column needs to be replaced. This example highlights that it is not the absolute column lifetime which has to be considered. Also, the sample throughput needs to be taken into account.

The question now arises as to which columns can be used for high-temperature operation? As has been noted in Chapter 1, silica-based materials were not sufficiently stable when the concept of high-temperature HPLC was introduced some decades ago. This was the reason why the group of Carr and co-workers has devoted much effort to synthesize alternative materials composed of zirconium dioxide which are coated with a polymer.[3,4] These metal oxide stationary phases have long become synonymous with high-temperature HPLC. Nowadays, these columns are commercially available and many more surface coatings have been developed since the early 1990s.

In addition to these metal oxide stationary phases, there are a lot of other materials which have also proven to be very well suited to high-temperature operation. These comprise purely polymeric materials like polystyrene–divinylbenzene or stationary phases completely composed of carbon, like the graphitized carbon column.[5–9] The main drawback of these stationary phases is that they are very hydrophobic or interactive and thus require a high proportion of the organic modifier in order to elute very retentive compounds. As will be shown in Chapter 8, there is an enormous potential for so-called "water only" separations. This can be either due to new detection systems which can only be used with water as the sole eluent, or a special method which requires that the content of the organic solvent in the mobile phase is very low. Hence, it is of great interest to reduce the amount of the organic co-solvent as much as possible and thus a stationary phase with a high hydrophobicity will not be the first choice for such applications.

In the last few years, however, column manufacturers have made tremendous efforts to increase the stability of reversed-phase columns based on silica. In the early days of high-temperature HPLC, some authors had already evaluated the temperature stability of commercially available silica-based reversed-phase columns, but the results were not very encouraging. Often such columns were degraded after only a few hours when the temperature was increased up to 100 °C. This has led to a widespread reluctance to have a closer look at this topic because such a rapid breakdown of the column is surely not acceptable. I recall that when I was doing my PhD thesis, I also tried a lot of different columns. Since the budget for university research is always scarce – in most cases you will even find that there is really no budget to perform your experiments – I searched through all the drawers in the whole laboratory. I was quite amazed that a large set of about 20 columns had been left after a project of a colleague was finished. Neither my colleague nor the system he had developed were there at the time I started my PhD. Therefore, I made my first experiments on high-temperature HPLC by exploiting the set of columns he had left for me. Unfortunately, I very soon realized that the large pool of columns would have been consumed after only a few weeks, because the columns rapidly degraded once the temperature was increased to 100 °C. This meant that I had to look for some alternative columns and these were not for free. Luckily, after some years of working quite intensely on this topic, I can make good suggestions about which columns can be used for high-temperature HPLC, so that the practitioner will not be faced with the same problems I faced when doing my PhD. *Although it is still thought that silica-based stationary phases are not suitable for high-temperature HPLC, with the advent of hybrid technology this notion is no longer valid. In addition to silica-based materials, coated metal oxide stationary phases, purely polymeric phases and materials made completely of graphitized carbon have long been used for high-temperature HPLC.*

2.3 The Detector

I have already briefly discussed the problem of cooling down the eluent before it enters the detector. The general notion is that a decrease in signal intensity will always occur if the temperature of the mobile phase entering the detector is increased. However, the user has to distinguish between two types of detectors. There are detectors where a high eluent temperature will lead to a worse performance. Also, an irreversible damage of the detector cell can result if the eluent temperature is too high. The fluorescent detector might serve as a good example. On the other hand, some detectors might profit from a higher eluent temperature. These comprise detectors where the eluent is transferred from the liquid to the gaseous state. The mass spectrometric detector, the evaporative light-scattering detector and the charged aerosol detector are based on the principle of a phase conversion of the eluent during the detection process. This means that a different approach is necessary when eluent cooling is considered. A severe problem may arise when the temperature of the mobile phase is above its boiling point. In this case, a phase transition may occur in the system. I will therefore close this chapter with some thoughts on back-pressure control.

2.4 The Back-Pressure Regulator

The back-pressure regulator is a simple device which keeps the pressure above a defined threshold provided that there is an eluent flow through the system. The question arises as to how to adjust the back pressure? As will be shown in Chapter 4, the vapour pressure of the mobile phase strongly depends on temperature. Without going into detail at this point, the highest vapour pressure of about 40 bar occurs if pure methanol is used at 200 °C. *In order to completely rule out a phase transition of the mobile phase somewhere in the system, the back pressure should be adjusted to at least 40 bar.* Now it must be decided where in the system to install the back-pressure regulator. Clearly band broadening should be as low as possible to maximize the efficiency of the chromatographic separation. Therefore, the best way to fulfil this requirement is to put the back-pressure regulator behind the detector. However, there are two reasons why this might not be feasible. Again, the type of detector has to be considered. The restrictor can be placed behind the detector if a UV, fluorescence or refractive index detector is used if the flow cell is sufficiently robust. This is not possible when a mass spectrometer, an evaporative light-scattering or a charged aerosol detector is used. In these cases, the restrictor has to be put before the detector and will therefore add to the extra column volume.

In order to minimize extra column effects, however, it is possible to use a restriction capillary instead of a back-pressure regulator. Furthermore, for those detectors where the mobile phase is transferred to a gas during the detection process, heating the mobile phase close to the transition temperature from the liquid to the gas phase might be advantageous in terms of detection sensitivity. Thus, cooling the mobile phase down to room temperature after it leaves the column is not necessary. This means that the hot eluent can be introduced into the detector. The problem is that depending on the flow rate of the mobile phase, the length and internal diameter of the restriction capillary, the composition of the mobile phase and the type of the organic modifier, a phase transition can occur in the connecting capillary. This means that the use of a restriction capillary, instead of a back-pressure regulator, requires some knowledge about the dependence of both the vapour pressure and the viscosity of the mobile phase on temperature. In liquid chromatography, the viscosity of the mobile phase is also a strong function of temperature and usually decreases with increasing temperature.

It is obvious as soon as we look more closely at the detail that more knowledge is required to design a high-temperature HPLC system. This chapter is meant to make the reader familiar with some general aspects of the system set-up for high-temperature HPLC. In the following chapters more specific information is given on these aspects.

References

1. S. Wiese and T. Teutenberg, unpublished results.
2. R. Agarwal, *J. Sep. Sci.*, 2008, **31**, 128.
3. M. P. Rigney, T. P. Weber and P. W. Carr, *J. Chromatogr.*, 1989, **484**, 273.
4. J. Nawrocki, M. P. Rigney, A. McCormick and P. W. Carr, *J. Chromatogr., A*, 1993, **657**, 229.
5. T. Yarita, R. Nakajima and M. Shibukawa, *Anal. Sci.*, 2003, **19**, 269.
6. M. M. Sanagi and H. H. See, *J. Liq. Chromatogr. Relat. Technol.*, 2005, **28**, 3065.
7. T. Teutenberg, O. Lerch, H.-J. Götze and P. Zinn, *Anal. Chem.*, 2001, **73**, 3896.
8. L. Pereira, *LCGC North Am.-Appl. Noteb.*, June 2006, 75.
9. L. Pereira, S. Aspey and H. Ritchie, *J. Sep. Sci.*, 2007, **30**, 1115.

CHAPTER 3
The Heating System

In this chapter, the focus is on the heating system. I have already briefly discussed the necessity for eluent preheating in Chapter 2. In this chapter an impressive example was given of peak distortion due to a complete lack of mobile phase preheating (see Figure 2.2). However, the knowledge that preheating of the mobile phase is essential when working at elevated temperatures is not new. In 1981 Abbott published a paper in which he already pointed out that high eluent temperatures will only help to improve the separation efficiency if adequate preheating of the incoming solvent is applied.[1] This means that without a specially designed heating system the advantages of high eluent temperatures cannot be put into practice. Although instrumental considerations highlighting the necessity of eluent preheating were published 30 years ago, the availability of dedicated heating systems for high-temperature operation was not actively pursued by most instrument manufacturers at that time. This can probably be attributed to the fact that silica-based stationary phases did not show sufficient long-term stability at elevated temperatures and hence there was no need to further improve the column oven. Nevertheless, much progress has been made in the last few years. However, there are different technical approaches to heating the column and the mobile phase. In the following paragraphs, a modular oven concept based on mobile phase preheating, column heating and mobile phase cooling will be considered and discussed. The chapter is then closed with a critical comparison between air-bath, water-jacket and contact heating ovens. Please note that I will not give an overview of all the heating systems which are commercially available. In the last few years new heating systems have been introduced to the market and I think that any survey would be rapidly outdated. What I consider important is that the reader is able to distinguish between the different technical approaches and is able to weigh the pros and cons of each system.

3.1 Preheating of the Mobile Phase

It is now generally recognized that eluent preheating is absolutely necessary to prevent the formation of axial and radial temperature gradients within the column.[2] This problem, which has also been termed "thermal mismatch", arises because silica or polystyrene column packings are relatively poor thermal conductors compared with stainless steel. However, discussions about the acceptable temperature difference between the incoming mobile phase and the stationary phase are in many cases rather academic with no practical benefit for the user. Most theoretical assumptions about the length of the preheating tubing are made on calculations which must be viewed with great care. Experimental evidence is often missing to support the conclusions based on these calculations.

3.1.1 Thermal Mismatch Broadening

First of all, let us consider what happens to the flow profile of the mobile phase inside the column if the eluent is not thermally equilibrated. A very nice illustration is provided by Wolcott and Snyder, which is also used here to highlight the effect of band broadening caused by thermal mismatch.[3] In Figure 3.1, an HPLC column is depicted which is heated to 70 °C. Figure 3.1a illustrates the situation when the mobile and stationary phases are both thermally equilibrated. In this case, no band broadening due to thermal mismatch is observed. The second example (Figure 3.1b) describes the problem when the mobile and stationary phases are not thermally equilibrated. Here, the eluent is not preheated to the temperature of the stationary phase and enters the column at a low temperature (*e.g.* 22 °C). Since only the column walls are heated, the solvent near the column wall heats up faster than the solvent in the centre of the column and a radial temperature gradient builds up. Consequently, the viscosity of the eluent along the wall is lower than that at the column centre and hence the mobile phase moves faster at the column wall than in the centre of the column. A convex flow profile is generated, causing the chromatographic band to broaden which may even result in split peaks. Thus, the overall separation quality and efficiency are both reduced. In addition to the radial temperature gradient, there is also an axial temperature gradient, because the temperature of the mobile phase gradually approaches the oven temperature at the column outlet. *Although the temperature can be accurately controlled with a precision of about $+/- 0.1 K$ for most commercially available heating ovens, this does not automatically mean that there is a homogenous temperature profile in an axial or radial direction within the column.*

Fields *et al.*[4] demonstrated the effect of using two different preheating coil sizes and mobile phase flow rates, ranging from 0.7 to 1.5 ml min^{-1}, on the peak shape of steroids. The authors observed that using a 15 cm long coil and a flow rate higher than 1 ml min^{-1} led to distorted or even split peaks when the temperature of the column was adjusted to 160 °C. By replacing the 15 cm long coil with a 150 cm long coil, peak shapes were symmetrical even at the highest flow

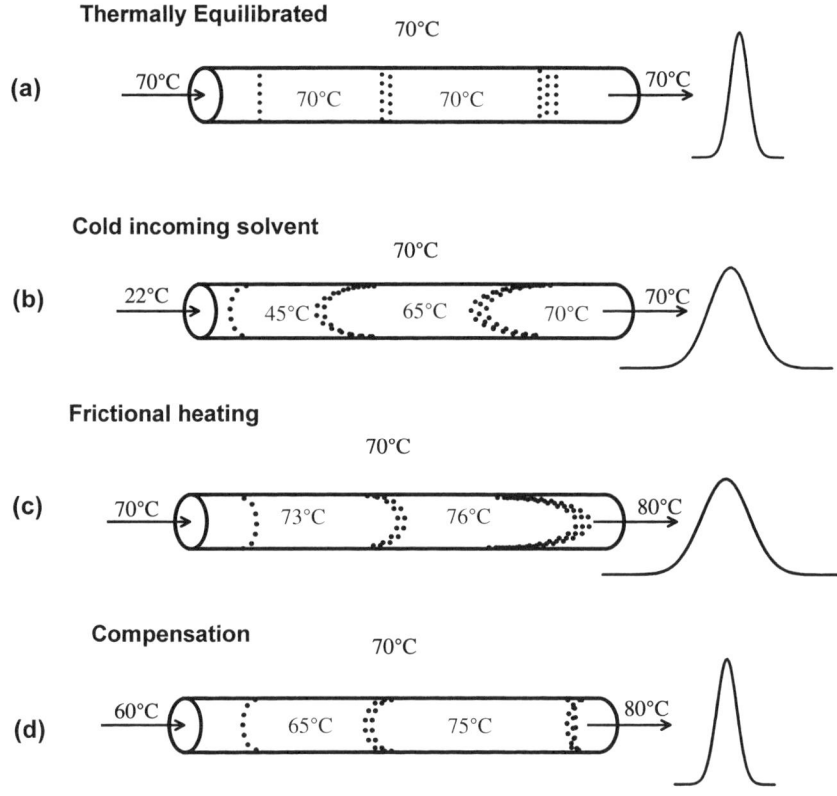

Figure 3.1 Band broadening due to thermal effects. (a) Ideal case, no thermal effects; (b) effect of incoming mobile phase that is at a lower temperature than the column; (c) effect of frictional heating; and (d) combined effects of cold incoming mobile phase and frictional heating. An oven temperature of 70 °C is assumed. Numbers shown inside column suggest plausible solvent temperature at column centre.[3] (Reproduced with kind permission from Elsevier.)

rate of 1.5 ml min^{-1} and a temperature of 160 °C. This example, which is given in Figure 3.2, again serves to illustrate the effect that a thermal mismatch between the incoming mobile phase and the stationary phase leads to severe band broadening or even peak splitting.

The question arises as to what is the required length to guarantee that the mobile and stationary phases are thermally equilibrated? Is it really necessary to use such a long preheating capillary as suggested by Fields? The answer depends on the heating system which is employed.

In a recent study, we have demonstrated the feasibility of a specially designed heating system for high-temperature gradient operation based on block heating.[5,6] At a temperature of 185 °C and a flow rate of 5 ml min^{-1}, a baseline separation of a mixture containing four steroids on a polybutadiene-coated

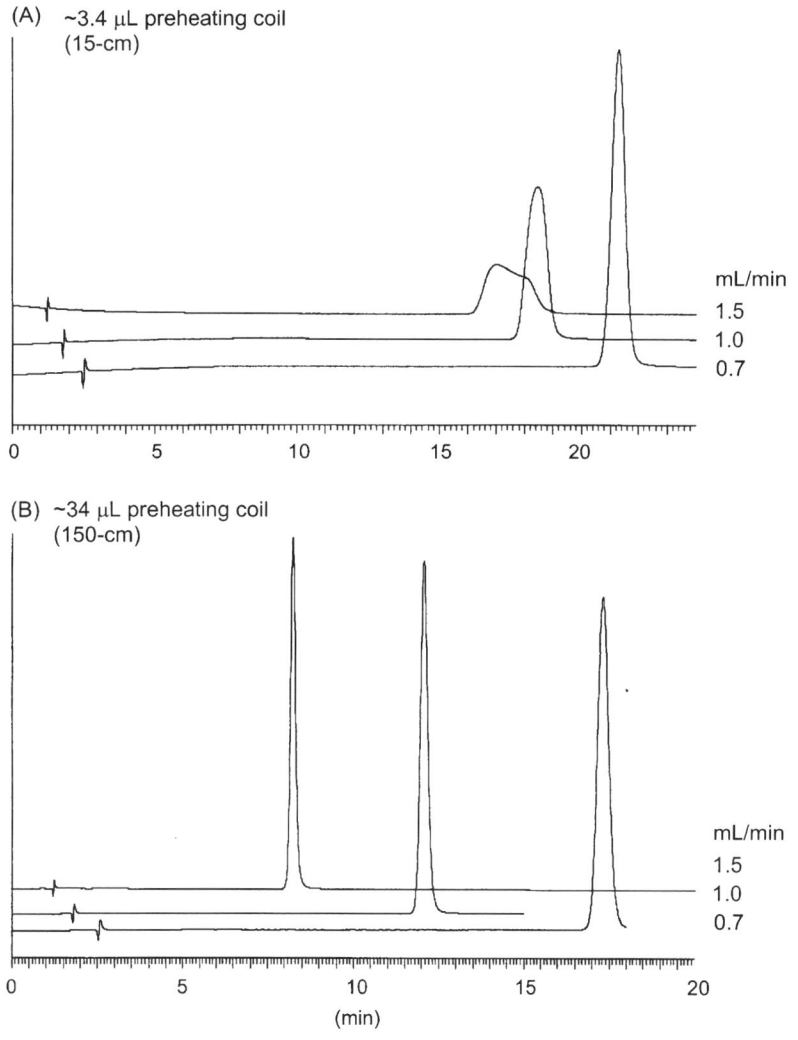

Figure 3.2 Effect of preheating coil length and flow rate on retention and peak shape. (Flow rates indicated on figure). Preheating coil volumes: (A) ~3.4 µl and (B) ~34 µl. Conditions: temp.: 160 °C; zirconia PBD column; 7 µl injection of testosterone; (A) 65 mAUFS and (B) 100 mAUFS.[4] (Reproduced with kind permission from Elsevier.)

zirconium dioxide column could be achieved with all peaks eluting symmetrically within 1.2 minutes. The length of the preheating capillary was only 15 cm. How can this discrepancy between the experimental results be explained? A reasonable explanation can be given when the systems which were used in these studies are compared. While Fields used a Gilson 831 column oven based on forced-air convection, we used a block-heating oven with the preheating capillary tightly clamped between two aluminium blocks. In addition, our experimental results are in good agreement with the mathematical treatment of Abbott and co-workers. They had already stressed the importance of eluent preheating in a paper published in 1981 about the effect of radial thermal gradients in elevated-temperature HPLC.[1] They calculated that for pure water as a mobile phase, which is pumped through a stainless-steel tube with an internal diameter of 0.23 mm at a flow rate of 2 ml min^{-1}, the temperature difference between the wall temperature and the temperature of the water is only about 1 °C if the water is heated from 20 °C to 70 °C in a 7.5 cm long tubing.

It can be summarized that heat transfer is most effective if contact heating is used. The mathematical treatment as well as the experimental evidence unambiguously suggest that it is not necessary to use an extremely long preheating capillary. If the heating system is well designed, a few centimetres will suffice. Hence, the fear that a preheating coil will significantly add to the extra-column volume is clearly overestimated.

Guillarme *et al.* presented a good example which highlighted that air is a very bad heat transfer medium.[7] They calculated the necessary length of the preheating capillary when an air-based oven was used and flow rates between 3 and 10 ml min^{-1} were applied. The results are given in Figure 3.3.

They suggested that for a flow rate of 5 ml min^{-1} at 200 °C, the preheating tube should be as long as two meters to reduce the temperature difference between the set oven temperature and the actual mobile phase temperature below 5 °C. They validated the theoretical assumption by comparing the efficiency for a separation of alkylbenzenes at a flow rate of 4 ml min^{-1} at 150 °C. There is no doubt that when a one meter long capillary was used, the late eluting peaks were severely distorted which was a clear sign of thermal mismatch band

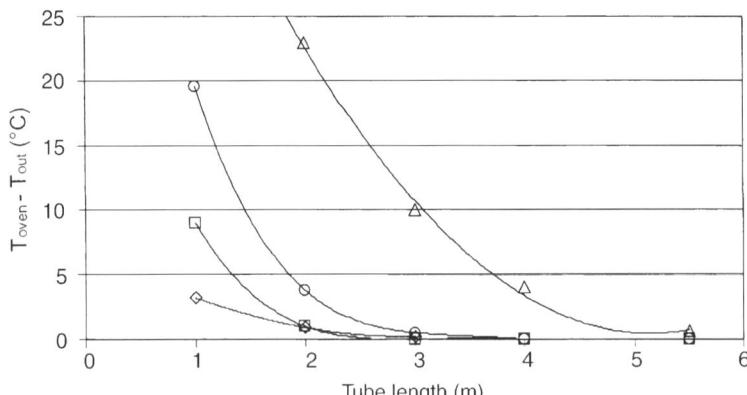

Figure 3.3 Experimentally measured differences in temperature between the incoming eluent and the oven as a function of the stainless-steel tube length for different oven temperatures and mobile phase flow rates: ◊, 100 °C and 3 ml min^{-1}; □, 150 °C and 4 ml min^{-1}; ○, 200 °C and 5 ml min^{-1}; and Δ, 200 °C and 10 ml min^{-1} (ref. 7). (Reproduced with kind permission from Elsevier.)

The Heating System

broadening. Replacing the one meter capillary by a two meter capillary helped to improve the peak profiles which can be seen in Figure 3.4. Please note that the calculation was based on an internal diameter of 127 μm. The conclusion that the preheating tubing adds significantly to the extra-column volume is therefore correct. However, as I have shown above, if contact heating is used instead of air, only 15 cm of a 127 μm ID capillary will bring the eluent to the desired temperature, even at flow rates up to 5 ml min^{-1}.

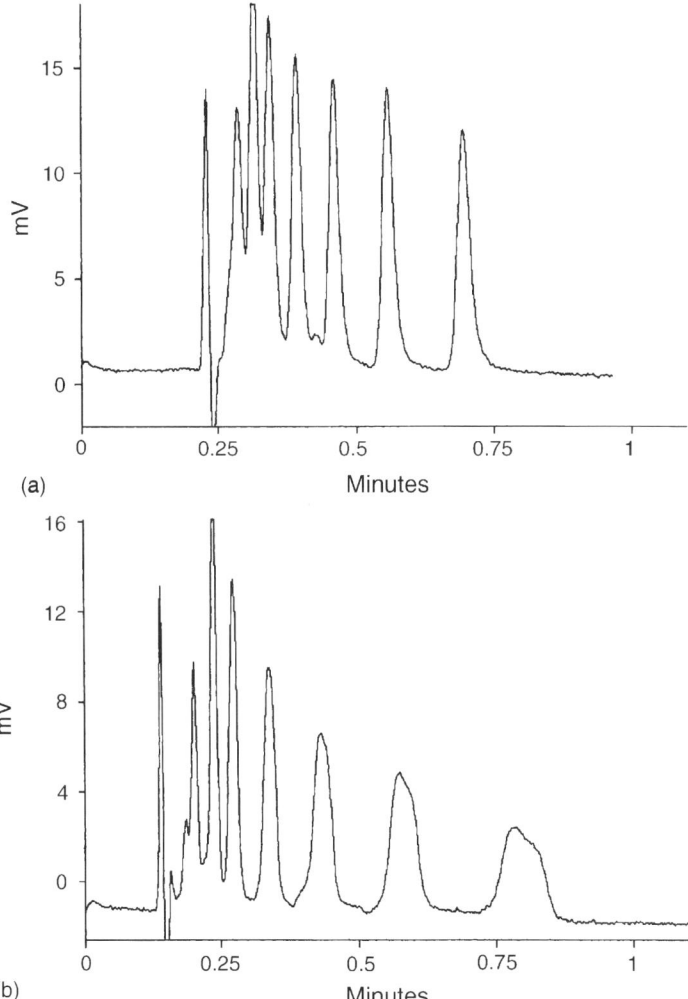

Figure 3.4 Chromatogram showing the effect of an efficient (a) and inefficient (b) preheating on the peak shape. Solutes: benzene, toluene, ethylbenzene, propylbenzene, butylbenzene and pentylbenzene. Chromatographic conditions: column: ZirChrom-DB-C18 (50 mm × 4.6 mm ID); flow rate: 4 ml min^{-1}; temp.: 150 °C; preheating tube: 127 μm ID with lengths of (a) 2 m and (b) 1 m.[7] (Reproduced with kind permission from Elsevier.)

In my opinion, this example clarifies that still air is a good insulator and heat transfer is very bad, which underlines the summary I have given above. I will resume this discussion in Chapter 3.1.4, where data is given on the experimental validation of the eluent preheating efficiency for contact heating.

Before I continue with the next paragraph, I will add a few words to the question of many practitioners regarding the benefit of thermostatting the solvent reservoirs, the pumps and the injection valve. In principle, thermal mismatch should be not an issue if all these modules were also thermostatted. However, there are some serious arguments which clearly speak against this approach. First of all, the technical requirements needed to thermostat the solvent reservoir, the pumps and the injector are much higher than just the design of an advanced heating system. Moreover, it is not helpful to thermostat the injection valve when the transfer capillary connecting the injector with the column is at ambient temperature. In this case, heat exchange would occur, leading to a significant cooling of the eluent before it reaches the column. A nice example which underlines this assumption has been given by Gritti and Guiochon.[8] In this publication, the authors pursued the question as to whether the mobile phase temperature at the column inlet would be raised due to a compression effect of the eluent in the pumps. Although the eluent is rapidly compressed by the pumps and the temperature at the pump exit is somewhat higher than the ambient temperature, they didn't notice an increase in the mobile phase's temperature at the column inlet. Hence, the best way to preheat the mobile phase is to use a short capillary in front of the column.

A second reason why not to thermostat the autosampler or the injection valve is that many practitioners are concerned about the degradation of analytes at elevated temperatures. Heating the autosampler would mean that the samples will be exposed to high temperatures for much longer. Additionally, if you would like to employ very high temperatures, the injection valve would need to be designed so that no phase transition occurs. This can cause problems if the normal boiling point of the sample solution is exceeded and it is not possible to control the back pressure.[i]

Nevertheless, the technical feasibility of thermostatting the complete HPLC system has been shown for gel-permeation chromatography (GPC). In GPC analysis, a widespread problem is being able to guarantee that the polymers which are being separated will not precipitate at any point in the system due to a reduced solubility at cold spots. This means that it is often necessary to work at very high temperatures using organic solvents like tetrahydrofuran. Here, temperature is predominantly used to keep the polymers in solution, so all connecting and transfer capillaries need to be thermostatted. A very impressive

---

[i] In a publication by Thompson and Carr, the authors came to the conclusion that with a maximum injector inlet temperature of 75 °C, there was almost no effect on the reduction of injector-to-column tubing length, assuming a desired column temperature of 200 °C. At higher flow rates, to see a 50% reduction in the length of injector-to-column tubing needed to heat the eluent to column temperature (200 °C), the eluent would have to be heated before the injector to greater than 170 °C. Consequently, no significant advantage would be obtained unless the injector could be operated at nearly the same temperature as the column.

system which is used for high-temperature gradient chromatography for the analysis of macromolecules and polymers has been developed at the German Institute for Polymers in collaboration with Polymer Laboratories. In this system, the entire instrument, from injector to detector, can be thermostatted so that a precipitation of polymers at low temperatures will not occur.[9,10] However, such a system would not be suitable for reversed-phase separations of small molecules, because the solubility of these molecules is not usually a problem at ambient temperature.

3.1.2 Viscous Heat Dissipation

If you now think that it is sufficient to simply adjust the temperature of the eluent to the temperature of the stationary phase, you are mistaken. In contrast to thermal mismatch band broadening due to an insufficient preheating of the mobile phase at very high temperatures, there is an adverse effect which might also lead to a reduced efficiency. This effect is called "viscous heat dissipation" or "frictional heating" and many authors have dealt with this phenomenon.[11–16] Consideration of this problem has become even more important in recent years due to the commercial availability of small particle packed columns, which often generate an extremely high back pressure. Nowadays, HPLC systems are readily available which allow for a maximum operating pressure of about 1000 to 1200 bars.[17–21] Interestingly, even as recently as ten years ago, it was always said that the pressure in an HPLC separation should be lower than 200 bars to reduce the wear and tear on the equipment. Today, the situation is completely different. Systems for very high-pressure operation have been successfully introduced into all the important fields of application. However, this means that in order to keep the loss of efficiency due to viscous heat dissipation to a minimum, a different strategy from that outlined in the previous section must be employed. But first of all I would like to recall what is meant by "viscous heat dissipation".

When a fluid is forced through the packed bed of the stationary phase by means of high pressure, the mechanical energy which was exerted by the pumps is converted into thermal energy.[ii] This thermal energy then leads to an increase in the temperature as the energy is dissipated in the stationary phase. Now it is important to differentiate between two possible ways of heat dissipation. If the

---

[ii] As a quantitative measure of frictional heating, the amount of power to be dissipated can be calculated using the equation: Power $= \Delta P \cdot F$, where ΔP is the pressure drop over the column in N m and F is the volume flow rate in $m^3 s^{-1}$. From this equation, several ways to minimize the formation of radial and axial temperature gradients in HPLC are evident. The first is to work at low pressure, since the power to be dissipated is directly proportional to the operating pressure. However, as mentioned above, a major trend in modern HPLC is to use small particle packed columns. In order to obtain fast separations, these columns are operated at high flow rates, which are not detrimental to the efficiency and thus, high pressures will be obtained. However, a very effective strategy is to reduce the diameter of the column. Since the optimal flow rate for the smaller diameter columns is lower, less frictional heating is generated. Also, a better heat dissipation can be achieved, preventing large radial temperature gradients. However, it should be noted that the instrument needs to be optimized in terms of extra-column volume. Otherwise, a low efficiency will be obtained due to band spreading in the connecting tubing.

column is assumed to be an adiabatic system, which would mean that it is totally insulated against the environment, heat dissipation through the column walls would not be possible. In this case, the heat can only be dissipated along the axis of the column. Considering that there is a pressure gradient along the column axis and that the mobile phase flows through the stationary phase, heat is transported from the head of the column (where there is the highest pressure) to the bottom of the column (where the pressure is lower) and gradually approaches atmospheric pressure. Consequently, an axial temperature gradient would build up in the column. Due to the fact that in a truly adiabatic system there is no chance to release heat *via* the surrounding walls, a radial temperature gradient would not be observed. If the temperature in the radial direction is constant, there would be no band broadening due to a difference in the viscosity either. However, in reality, the column is not completely insulated from its environment. This means that there is always a dissipation of energy through the column walls. Depending on the way the column is heated or thermostatted, heat can be readily dissipated. This is the case if, for example, a water bath is used as the thermostatting medium. Water has a higher thermal conductivity than still air and heat can be transported away from the column wall. Some authors have denoted this "a well thermostatted" column, because the heating medium has a high thermal conductivity and is in close contact with the column walls.

What does this now mean for the flow profile which will be observed if heat is generated inside the column due to frictional heating? In principle, a flow profile as depicted in Figure 3.1c will result. If we assume that the eluent is adequately preheated and is brought to the same temperature of the stationary phase, an increase in the column temperature due to frictional heating would occur at high operating pressures. If, in this case, the column is thermostatted under adiabatic conditions, the heat cannot be dissipated *via* the column wall. Rather, viscous heat dissipation leads to the formation of a longitudinal temperature gradient instead of a radial temperature gradient. Although a truly adiabatic environment is hardly ever encountered under standard HPLC operating conditions, this situation is approached to a certain extent when still air ovens are used to regulate the column temperature, because still air is a very good insulator. If a "well thermostatted" column is used which is the case when the column is heated by a water-bath or contact heating, heat dissipation into the surrounding medium is possible and thus, a radial temperature gradient will be generated. This radial temperature gradient now has the same effect as described in section 3.1.1, with the difference being that the resulting flow profile of the mobile phase is inverted (compare Figures 3.1b and 3.1c). In this case it might appear that still air ovens would be ideally suited if one likes to work at very high pressures, although this kind of oven has a poor heat transfer. A much more intelligent approach is to pre-cool the mobile phase to a temperature below the adjusted column temperature to avoid band broadening (see Figure 3.1d). Although this sounds very contradictory at first sight, it is absolutely reasonable and there is also experimental evidence that a pre-cooling of the eluent has a positive effect to minimize a loss in efficiency.

The effect of band broadening due to frictional heating has been predicted theoretically and verified experimentally by the early work of Poppe and Kraak, who were able to show that the solute band moving through the column will be distorted more when a column is operated at room temperature and a high flow rate is applied.[11,22] In one of these experiments, a 15 cm × 4.6 mm ID column was run with a mixture of a 31 % aqueous ethanol solution, at 30 °C. The flow rate was adjusted to 1, 2 and 3 ml min^{-1} yielding a pressure of about 100, 200 and 300 bars, respectively. The feed temperature of the eluent was varied from 16 to 50 °C. While at a low flow rate of about 1 ml min^{-1} there was no influence of the eluent feed temperature on band broadening, at higher flow rates and thus higher pressures a steep increase in the minimum plate height could be observed for feed temperatures higher than 25 °C.

In a more recent study, Mayr and Welsch demonstrated that it is possible to maintain a high efficiency by pre-cooling the mobile phase when a column was operated around ambient temperature and high pressure.[12] The authors used different columns packed with non-porous 1.5 µm particles which were placed in a liquid thermostat. In order to discuss the results I have adapted the chromatograms from their study, which are depicted in Figure 3.5.

The chromatograms were all recorded at the same column temperature, which was adjusted to 30 °C. The different traces resulted when the eluent temperature was varied between 5 °C (trace a) and 30 °C (trace d). Obviously, the best column performance was observed if the eluent temperature was adjusted to 10 °C. A lower eluent temperature favoured the formation of peak shoulders, especially for the highly retained compounds 5 and 6, while higher eluent temperatures also led to severely distorted peaks. Interestingly, if the eluent temperature exactly matched the column wall temperature, split peaks were observed and the efficiency was lowest. This example clearly highlights that when the column is operated at a "low" temperature but high pressure, pre-cooling the eluent will reduce the formation of radial temperature gradients so that the efficiency can be maintained. In this instance, the two effects which lead to radial temperature gradients counterbalance each other.

Besides the phenomenon of peak splitting, another effect visible in the chromatogram concerned the retention of the solutes. Please note that the column was always operated at the same temperature and only the inlet temperature of the eluent was changed. What is usually totally neglected is that retention of the analytes will be also affected if the eluent temperature varies, although the column temperature is kept constant. In order to avoid any misunderstandings, I have to stress that the term "eluent temperature" here means the temperature of the eluent as it enters the column. This has to be distinguished from the "column temperature" which is the temperature of the surrounding medium of the column and does not necessarily correspond with the temperature inside the column. In Figure 3.5, the effect of shifting retention times or retention factors is visible. The higher the eluent temperature, the lower was the retention of the solutes. Clearly, the peak movement was more pronounced for the more highly retained compounds. This explanation is quite straightforward, as a higher eluent temperature also leads to a higher

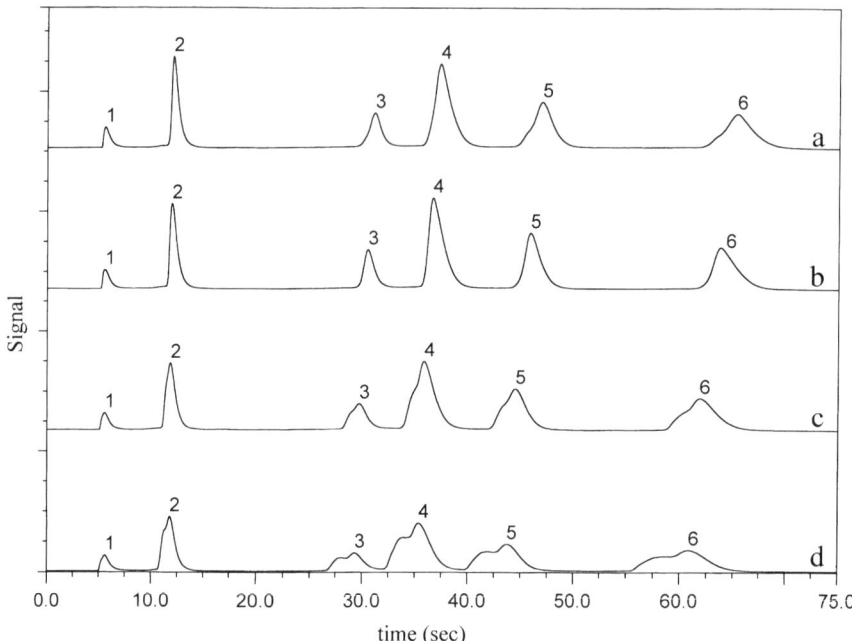

Figure 3.5 Separation of five hormones at different eluent temperatures. Column temp.: 30 °C. Trace a: eluent temp. 5 °C, pressure drop 508 bar; trace b: eluent temp. 10 °C, pressure drop 503 bar; trace c: eluent temp. 22 °C, pressure drop 490 bar; and trace d: eluent temp. 30 °C, pressure drop 480 bar. Mobile phase: acetonitrile–water (20 : 80 v/v). Flow rate: 2.0 ml min^{-1}. Injection vol.: 0.5 μl.[12] (Reproduced with kind permission from Elsevier.)

average temperature in the longitudinal direction of the column. For most of the small molecules which are separated in reversed-phase mode, increasing the column temperature leads to decreasing retention times. This means that one should carefully consider the length of the preheating tubing when a robust method is developed. Small changes in the overall length of the preheating tubing might result in a shift in the retention time of solutes, although the column temperature is kept constant. Ideally, if there are no axial temperature gradients in the column, the retention factor should be constant if the flow rate is increased. However, if on changing the flow rate axial temperature gradients are created due to frictional heating or insufficient preheating of the mobile phase, they can result in a lower or higher average temperature. Consequently, the retention factor may be a function of the flow rate, decreasing in the case of frictional heating or increasing with flow rate in the case of insufficient preheating at otherwise constant chromatographic conditions. The higher the flow rate, the more pronounced this effect will be.

What does this mean for the design of a heating system which is not only to be used for high-temperature HPLC but also for "conventional" and even ultra high-pressure applications? In my opinion, the system should be modular, which means that the eluent temperature can be controlled independently from the column wall temperature. Moreover, it should also be possible to adjust a lower or a higher eluent temperature than that of the column. In this case, the user has the highest degree of flexibility, regardless of whether the separation is carried out at elevated temperature and "low" pressure or *vice versa*. Although de Villiers *et al.* also commented on viscous heat dissipation,[14] they came to a different recommendation, which I would not fully support. While they acknowledged that temperature compensation by pre-cooling of the mobile phase was feasible to compensate band broadening, they stated that a practical implementation was not feasible in most routine laboratories. I do not agree with this statement because it would also mean that eluent preheating is not an option for routine analysis when working at high eluent temperatures, although it is generally recognized that this is the only way to maintain high efficiency at high eluent temperatures. Even Poppe and Kraak concluded in a very early paper that the suitable adjustment of the feed temperature can counteract band broadening and prevent a loss in efficiency due to viscous heat dissipation.[22] I think that the problem is more fundamental, because what is really needed is an "intelligent" heating device which is able to adjust the eluent temperature according to the operating conditions, so that there is no loss in efficiency due to radial temperature gradients. Then, the operator would not have to worry about which effect dominates under the conditions of the applied method. Unfortunately, it must be acknowledged that even if the heating oven allows for a separate adjustment of the eluent and column temperature, it requires "trial and error" to find the best conditions. From a practical perspective I would always recommended that the operator should critically evaluate the resulting chromatograms and vary the eluent temperature according to the operating conditions. *If very high temperatures are applied, the eluent temperature should be close to the column wall temperature because it can be assumed that viscous heat dissipation will play a minor role. If the column is instead operated at a low temperature but very high pressure, pre-cooling of the eluent might prove successful.* I have to stress that up until now no "intelligent" heating device exists which takes all variables into account and adjusts the eluent temperature accordingly. At this point more effort has to be devoted to put such a system into practice. However, with some of the most advanced heating systems currently on the market, this aim could be achieved in the near future.

3.1.3 Technical Implementation of Eluent Preheating

There are many ways to carry out eluent preheating and I would like to stress that it is not possible to consider all technical solutions. Instead, I will focus on some inventions which are also used in commercially available heating systems. Before I come to the more sophisticated devices, I will comment on the most commonly used systems as a starting point. Air ovens are probably still the most widely used devices for column thermostatting. However, as I have shown in the previous paragraph, these devices usually have no option for eluent preheating. This means that there is an undefined length of capillary which connects the column with the injector. If you are a good chromatographer who is concerned to reduce the extra-column dead volume as much as possible, you will probably use the shortest connecting capillary, but this might backfire because you have absolutely no opportunity to control the eluent temperature. If high temperatures are used you will probably have insufficient preheating. So this is definitely the worst option, and you will not be able to maximize the efficiency if you plan to increase the temperature or the pressure. Therefore, the column thermostat should also include a defined eluent preheating. One option is to use a device containing a short length of tubing with a heating or cooling element that is in close thermal contact with the exterior of the tubing. Although capillaries made of polyetheretherketone (PEEK) have long been used in HPLC (as they can be easily handled and used to make finger-tight connections without the need for additional tools), they are now being replaced by traditional stainless-steel capillaries because these are also pressure resistant up to 1000 bars or higher. It is clear that in order to achieve a fast and precise preheating of the eluent, only materials with a good thermal conductivity should be used. This will also help to greatly reduce the length of the preheating capillary required to bring the eluent to the adjusted oven temperature.

Figure 3.6 depicts the design of an eluent preheater based on contact heating. Here, the transfer capillary leading from the injector to the column is tightly enclosed between two aluminium shells. Due to the close contact between the aluminium block and the stainless-steel capillary, a fast and precise preheating

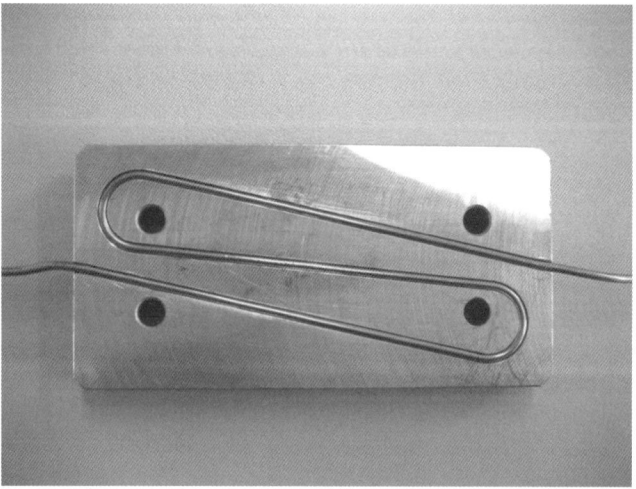

Figure 3.6 Eluent preheating based on contact heating. The capillary is tightly enclosed between two aluminium shells.

of the eluent is possible. However, there is another advantage of this set-up. As is evident from Figure 3.6, the length of the preheating capillary is precisely defined. This guarantees that if the capillary has to be replaced, the length of the preheating capillary will remain constant. This is a decisive advantage if a method transfer has to be made to a different HPLC system. Provided that the same column oven is used and the connecting capillary is kept constant, no change of retention times should be observed. This is completely different from conventional air-based column ovens where the length of the connecting capillary is not controlled. In such a case, the method may have to be adjusted in order to compensate for the shift in retention times.

A different approach to eluent preheating would be to immerse the capillary and the column in a liquid bath. If the temperature is not raised beyond 100 °C, a water bath could theoretically be used, but if you would like to extend the temperature range further, an oil bath would have to be used instead. While the preheating can be achieved very efficiently by a fluid, the fluid may be easily contaminated with the mobile phase if there is a leakage, in addition there is the danger that the solvents might boil out of the fluid. Since toxic organic co-solvents are used, this would mean that the fluid would have to be completely replaced once it has been contaminated. This procedure is not only very time consuming, but the more troublesome effect is that the system cannot be used for analysis while the fluid is being changed. Unfortunately, a leakage cannot always be avoided, especially if a new column is installed and the fittings are not properly connected. Moreover, separate control of the eluent temperature is not possible, as I will show in section 3.2.2.

To conclude this section both from a practical perspective as well as from a technical standpoint, a device which is based on contact heating seems to be the best option in order to allow for a rapid and precise preheating of the eluent. Ideally, this device should be controlled independently from the column so that efficiency can always be maximized, depending on the requirements of the method.

3.1.4 Experimental Verification of Eluent Preheating Efficiency

Before I proceed to the issue of column heating, I would like to give some experimental evidence which can be taken as a reference for the validation of the concepts I have described in the previous paragraph. These results have been published elsewhere,[5,6] but it seems appropriate to point out the most important elements and to make a comparison with other studies.

When we started with the development of a specially designed heating system some years ago, we were looking for an experimental verification of the preheating efficiency when block heating is used, *i.e.* when the capillary is brought into close contact with a metal heating block. Basically, the results were based on the measurement of the eluent temperature within a stainless-steel capillary. For this experiment, a thermocouple with an outer diameter of 0.15 mm was placed inside a stainless-steel capillary with an internal diameter of 0.25 mm. By introducing the thermocouple into the capillary we were able to record the temperature of the fluid inside the tubing. Theoretically, a mathematical treatment of the heat exchange was possible but when the diameter of the capillary was reduced to below 1 mm, it would have been a difficult task to apply the correct assumptions for heat transfer. The capillary was soldered into a copper block so that a good heat transfer was achieved. The thermocouple could then be placed at any point within the capillary, so that the temperature at the point where the eluent enters and leaves the heating block could be measured. Figure 3.7 depicts the measuring positions inside the capillary. Please note that the total length of the section which was inside the block was only 13 cm.

The first experiment served to confirm that the temperature of the eluent could reach that of the heating block over a length of 13 cm. As can be observed from Figure 3.8, at the highest block temperature of 190 °C the temperature difference between the adjusted block temperature and the eluent temperature was less than 1 °C at the outlet of the capillary. The flow rate of the eluent, which consisted of pure water, was adjusted to 2 ml min^{-1}. This is a very high flow rate considering the steady trend towards miniaturization of the column and particle diameter and that mass spectrometry is increasingly used for detection.

The second experiment was designed to evaluate how much the flow rate of the eluent could be increased without exceeding a temperature difference of 5 °C. The results are displayed in Figure 3.9, where the temperature differences of the adjusted block temperature and the eluent are plotted against the flow rate of the mobile phase. It is obvious that only after increasing the flow rate above 5 ml min^{-1} at a block temperature of 150 °C was the temperature difference greater than 5 °C.

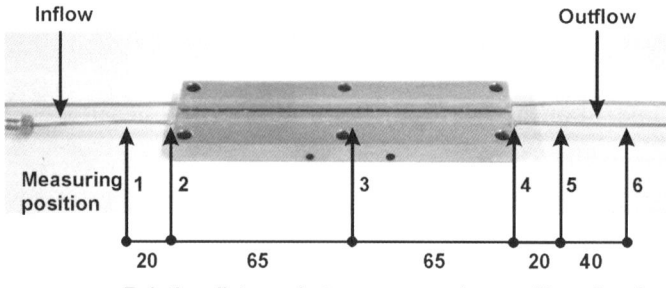

Figure 3.7 Experimental design for measuring eluent temperature in a stainless-steel capillary. Internal diameter of capillary: 250 µm; outer diameter of thermocouple: 150 µm. The thermocouple was placed at the depicted measuring positions inside the capillary. The stainless-steel capillary was soldered in a copper block of 130 mm length which was mounted on an aluminium heating block of the same length.[5] (Reproduced with kind permission from Elsevier.)

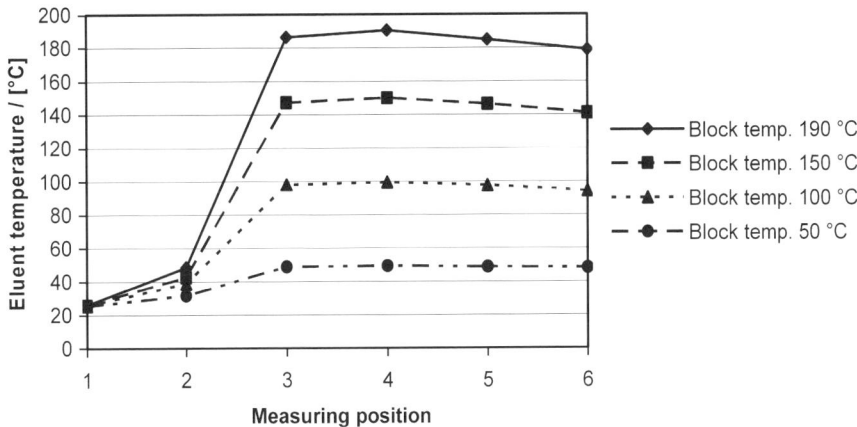

Figure 3.8 Dependence of measured eluent temperature in a stainless-steel capillary on different measuring positions at a constant flow rate of 2 ml min^{-1}. Deionized water was used as the eluent.[5] (For details of experimental design, see Figure 3.7.)

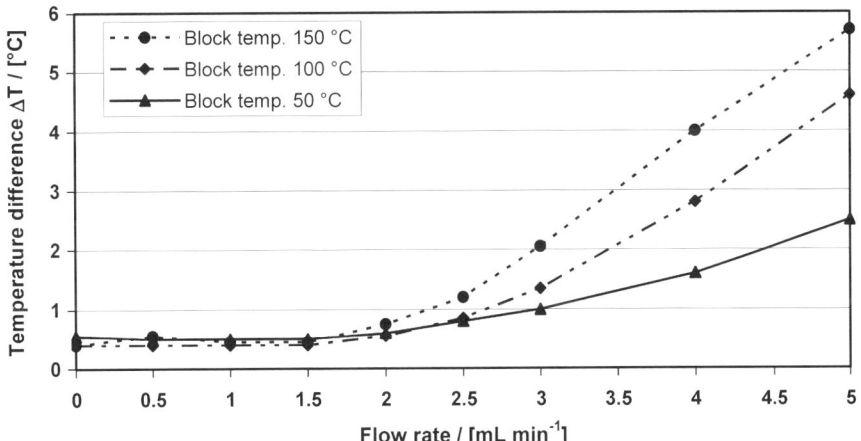

Figure 3.9 Temperature difference of measured eluent temperature in a stainless-steel capillary *versus* flow rate.[5] Temperature was measured at position 4. (For details of experimental design, see Figure 3.7.)

These results serve as an experimental validation and underline that the mathematical calculations made by Abbott can be met experimentally. When the results are compared to the experimental validation of Guillarme *et al.* presented in Chapter 3.1.1, it is obvious that contact heating is better suited to the rapid preheating of the mobile phase within a very short distance. Hence, extra-column band broadening can be significantly reduced.

3.2 Column Heating
3.2.1 Air-Bath Ovens

There are different approaches to heating the mobile and stationary phases. Probably the most common heating ovens for liquid chromatography are based on the same principle as GC ovens. In such ovens, air is used to heat the column to the desired temperature. Again, I would like to emphasize that the heat transfer of air is very poor when compared to a contact medium like water, silicone oil or a heated metal block. The reason is that air is usually a good insulator, especially if the air is not blown against the surface it should be controlling. While air-bath ovens might work well if the column heater is only used for maintaining a constant temperature, long equilibration times result if the temperature needs to be changed very rapidly for *e. g.* temperature programming. However, long equilibration times become not only critical when the temperature is increased to 200 °C, but also when the temperature is raised only slightly above the ambient temperature. In a recent study by de Villiers and Sandra, the authors noted that using the Acquity column heater, which is integrated into the UPLC^{TM} system of Waters, a thermal equilibration time of between 30 and 60 minutes was observed when the temperature was set to 40 °C.[14] The authors noted that after 20 minutes equilibration time the temperature was at 36.9 °C, while at 60 minutes the temperature reached 38.6 °C. After 60 minutes, no further increase in the temperature could be measured. This means that it took a long time until the system was in full thermal equilibrium and that the set temperature was not completely reached. Given the fact that these are moderate temperatures, it would take much longer when a substantially higher temperature is employed. This example highlights that still air is not a good medium for a fast heating of an HPLC column.

3.2.2 Water-Jacket Ovens

A water-jacket oven is much more efficient than an air-bath oven when heat transfer is considered. A clear advantage is that the fluid medium is in close contact with the column wall, thereby enabling a fast heat transfer. However, there are technical difficulties which have to be considered. In a recent research project where we developed a specially designed heating system for high-temperature liquid chromatography, we also tried to consider a system where the column and the capillary were immersed in a liquid bath.[6] A schematic drawing is given in Figure 3.10.

Here, the column was placed in a closed reservoir, which was thermally insulated. The preheating of the mobile phase was carried out by placing the capillary connecting the injector to the column in the liquid bath alongside the column. The column inlet was therefore at the bottom of the reservoir, which was sealed with a cap. In this cap there were holes to connect the capillaries with the column and detector, respectively. Although this approach would be suitable for high-temperature operation, we did not pursue this concept further

The Heating System

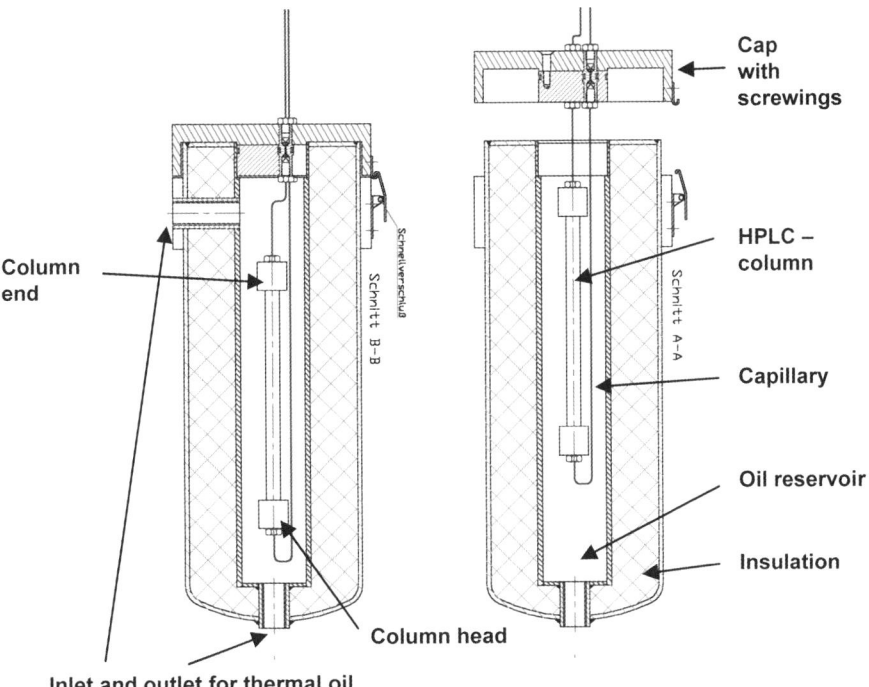

Figure 3.10 Scheme of a heating system where the column and the capillary are both immersed into a heating liquid.

because we deemed the application of such a system impractical in a routine environment. Another serious issue is that the silicone oil can be contaminated with the mobile phase in cases of leakage.

3.2.3 Block-Heating Ovens

A third option for heating the mobile and stationary phases is to use a contact heater. As has already been outlined, the heat transfer can be carried out most effectively by block heating, because a tight contact is established between the column and the heating unit. Figure 3.11 depicts the technical approach to column heating. The column is enclosed between two aluminium shells which are connected with a heating block. In order to optimize heat transfer, the aluminium shells have to be tailor made for each column.

An intimate contact must be guaranteed between the shells and the column, otherwise the formation of small voids can lead to local temperature differences in the radial and axial direction of the column. Also, the endfittings of the column have to be included. The problem is that due to the great inventiveness of column manufacturers, many different column designs have been created. Consequently, the shells for columns which nominally have the same size,

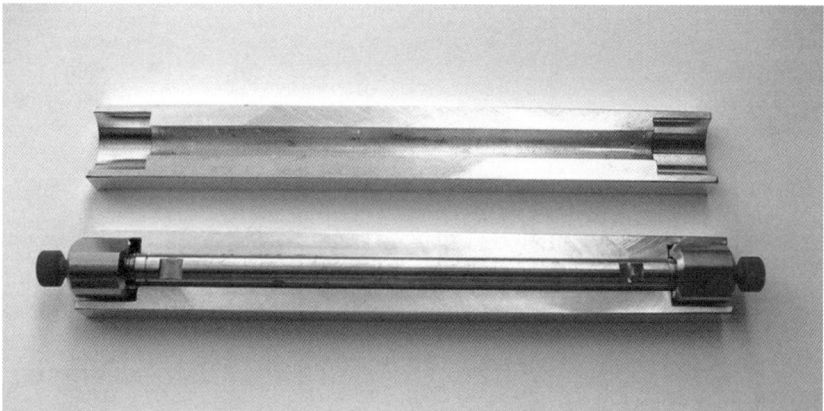

Figure 3.11 Column heating is realized by enclosing the column with two aluminium shells. These shells also enclose the column's end fittings.

length and internal diameter cannot be used for columns from different vendors. A unified column hardware would therefore be of great help, but I don't think that different manufacturers will agree on one common hardware concept. Therefore, the column shells have to be tailor made if a column from a different vendor is used, even if the column dimensions are virtually identical. Also, the shells have to be changed if two columns from the same vendor but of different length are used.

However, not only does the column temperature need to be controlled, but also the temperature of the eluent after it leaves the column, which will be discussed in the next section.

3.3 Post-Column Cooling of the Mobile Phase

Heating the mobile and stationary phases up to temperatures of 200 °C means that the eluent has to be cooled down prior to detection. This is also a serious requirement if special hyphenation techniques are applied, which I will discuss in Chapter 8. In most publications a laboratory-scale or home-made construction is used to cool down the mobile phase. A water bath or ice bath is placed before the detector in which the capillary is immersed. However, for routine analysis the temperature has to be controlled more accurately and this approach cannot be recommended, because the ice bath has to be replaced frequently. Therefore, the device for cooling down the mobile phase has to be integrated into the heating system. Modern heating systems should therefore allow for precise eluent cooling before the mobile phase is introduced into the detector. In such systems the capillary is also placed between two aluminium shells according to Figure 3.6 and a Peltier element is used for eluent cooling.

The need for a constant temperature prior to detection is highlighted by the next experiment, which should give valuable information about the deviations

The Heating System   43

in peak area as a function of temperature. In this experiment, a preheating capillary was connected directly to a UV detector without using a column. Then the peak area of test solutes was monitored. As can be seen from the results in Figure 3.12, increasing the temperature caused a decrease in the peak area of the analytes. Polycyclic aromatic hydrocarbons were chosen because the extraordinary stability of these compounds at high temperatures is well known. Furthermore, the residence time of the analytes in the heated capillary was just a few seconds. Therefore, a possible degradation of the compounds can be ruled out.

This example highlights that a precise control of the detector inlet temperature is mandatory. Although many detectors are equipped with thermostatted cells, the mobile phase will not be cooled down to ambient temperature if the eluent enters the detector at a very high temperature. I would like to point out that it is not possible to deduce a general rule based on the example given above. We have not only observed a decrease in signal intensity with increasing temperature, but also noticed that higher temperatures can have a positive effect on signal intensity. Again, a distinction should be made between the detectors where the eluent remains a liquid during the detection process and the detectors where the eluent is converted to a vapour. The evaporative light-scattering detector, the charged aerosol detector and the mass spectrometer can be assigned to the latter group. At the moment, the data is not sufficiently extensive to draw up general rules on the adjustment of the detector inlet temperature to a certain value. The practitioner has to keep in mind that even the temperature of the mobile phase after it has left the column plays an important role in the overall method optimization, exerting an influence on the signal-to-noise ratio or the peak area. Sometimes it is difficult to distinguish between effects where a degradation of analytes takes place on the column, or where the signal is diminished by temperature effects related to the detection process. I will resume this discussion in Chapters 6 and 7. *At the moment, it can be summarized that it appears to be useful to cool down the eluent temperature close to room temperature if a UV, fluorescence or refractive index detector is used. In contrast to this it might be helpful to introduce the mobile phase at a higher temperature for the second group of detectors, where a conversion of the mobile phase from a liquid to a vapour takes place.*

3.4 Temperature Programming

A much more demanding approach is to use temperature gradients in liquid chromatography. I will explain the necessity for using this technique in Chapter 6, where the focus is on method development strategies. Moreover, I will give examples based on new technologies and methods in Chapter 8, which rely on temperature programming. However, before I deal with applications where temperature programming is a prerequisite, I would like to explain how a temperature gradient works in contrast to a solvent gradient. Usually if the temperature is increased during a chromatographic run, the same effect on

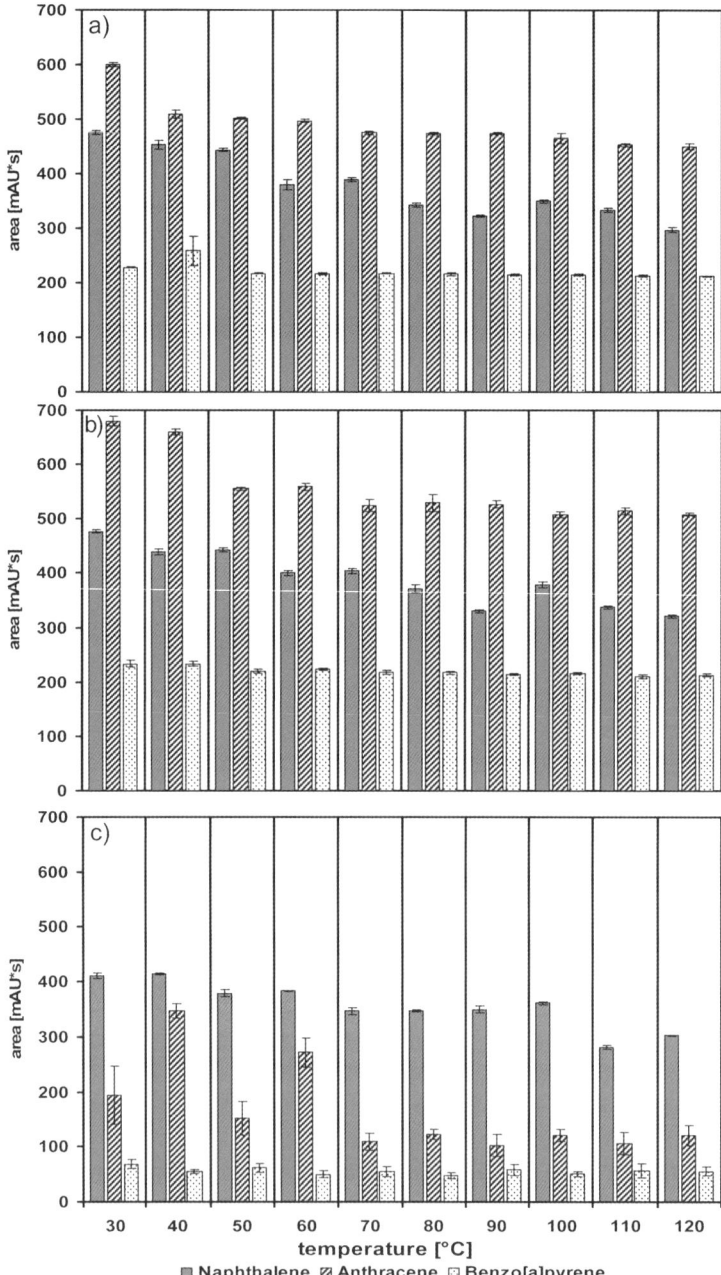

Figure 3.12 Peak area of three PAH at different eluent temperatures. Chromatographic conditions: solvent A: deionised water (containing 0.1% formic acid); solvent B: acetonitrile (containing 0.1% formic acid); flow rate: 0.5 ml min^{-1}; injection vol.: 1 µl; detection: UV DAD at 254 nm, with a fixed cell temp. of 30 °C. Eluent composition: (a) 100% solvent B, (b) solvent A–solvent B (50 : 50%) and (c) 100% solvent A.[25] (Reproduced with kind permission from Elsevier.)

analyte retention is observed as for the solvent-gradient mode, which means that the elution strength of the mobile phase is increased. The difference between these two gradient modes is that the solvent front always moves from the inlet to the outlet of the column, while temperature immediately affects the whole column. What needs to be considered is that, depending on the type of column oven, a significant lag may occur between the oven temperature and the programmed gradient.[iii]

The temperature delay is significantly dependent on the heating system and should be as small as possible. This delay factor is actually composed of two terms. One describes the temperature delay between the programmed gradient and the actual oven temperature. The second term describes the time it takes to heat up the stationary phase in the radial direction, because the heat has to be transferred from the column walls to the column centre. It is clear that the temperature inside an HPLC column is not uniform. No matter how efficiently the desired temperature is imposed on a chromatographic column, thermal gradients prevail inside the column. Reducing the inner diameter of the column is certainly the best way to get rid of this problem and temperature programming should be ideally suited for capillary and nano columns. Unfortunately, capillary or nano HPLC is not that widely used in industry up until now and hence I will concentrate on columns with an internal diameter of about 2 to 4.6 millimetres, which are by far the most frequently used column formats at the time of writing.

I would now like to describe an experiment where we have documented the efficiency of a heating system based on contact heating in temperature-programmed mode. The goal of this experiment was to measure the temperature inside an HPLC column with an internal diameter of 4.6 mm. This experiment should answer the question as to how pronounced is the lag in temperature of the packing material behind the oven temperature, when the temperature is increased during a chromatographic run. A column based on polybutadiene-coated zirconium dioxide was used and a thermocouple with an internal diameter of 0.15 mm was introduced into the column. The thermocouple was placed in the centre of the column packing and the column was enclosed by two tailor made copper shells. The heating rate was adjusted to $22\,°C\,min^{-1}$ and the temperature difference between the heating block and the inside of the column was recorded. We were able to show that up to a temperature of $90\,°C$, the temperature difference between the stationary phase and the block heater was less than $1\,°C$ (Figure 3.13). This means that an efficient heat transfer can be achieved even if the system is operated in temperature-programmed mode.

A possible drawback of block-heating ovens, however, might be that they act as a heat sink with a large thermal mass. When a separation is carried out under isothermal conditions, this is not a problem because the temperature is kept constant. However, the concept of using temperature programming must also include a fast cooling of the column to the starting temperature after the

---

[iii] In fact, this is also the same as in solvent gradient programming where the dwell time has to be considered, which depends on the mixing chamber and the pump system. The dwell time is defined as the time it takes for the mobile phase to travel from the point where the solvents are mixed, to the point where the mobile phase reaches the head of the column.

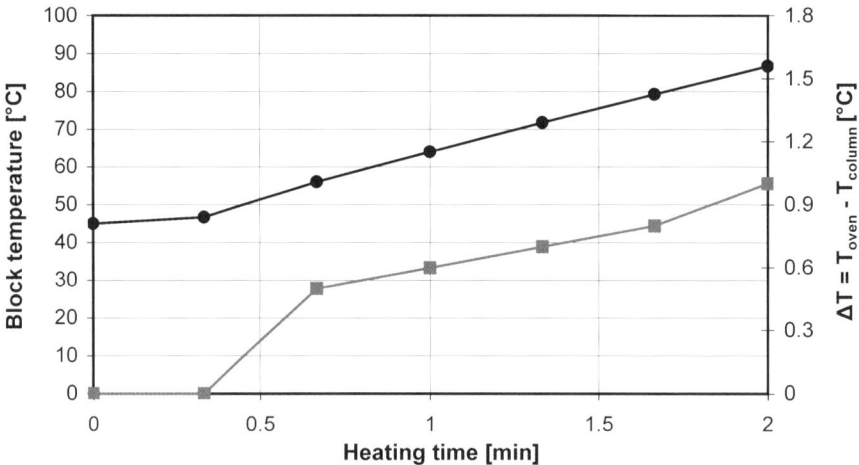

Figure 3.13 Dependence of block temperature, as well as temperature difference between block temperature and inner column temperature, on heating rate. The upper line represents the measured block temperature as a function of the heating rate. The lower line represents the difference between the measured temperature of the stationary phase within the column and the block temperature.

temperature gradient is finished. Although it is often assumed that a block-heating device is not suitable for temperature programming, I will demonstrate that the use of a block-heating oven can be very effective in terms of a fast temperature change after a temperature gradient. The idea behind this concept was to remove the heat from the column by applying a water cooling device which was placed behind the heating block of the column. After a temperature gradient was finished, the heat was removed by a flow of tap water. The effectiveness of such a cooling mechanism is demonstrated in Figure 3.14.

It can be seen that the column was cooled down from 150 °C to 50 °C in approximately two minutes. However, for this cooling process, 4.2 litres of water were consumed. It is possible to reduce this amount of water, but then it takes longer to reach the starting temperature. The amount of water can be reduced approximately four-fold from 4.2 to $1.2\,l\,min^{-1}$, while the time to cool down the column only increases by a factor of 1.25. Today this system has been brought to market and the cooling is achieved by using an internal cooling cycle with a refrigeration compressor.[23] This design means that the heating system is now completely independent of tap water.

3.5 A Critical Comparison between Different Ovens

3.5.1 Air-Based Ovens

If you have a close look at the ovens which are used in your laboratory, then I'm sure that air-based column ovens will be used predominantly. Why are

The Heating System

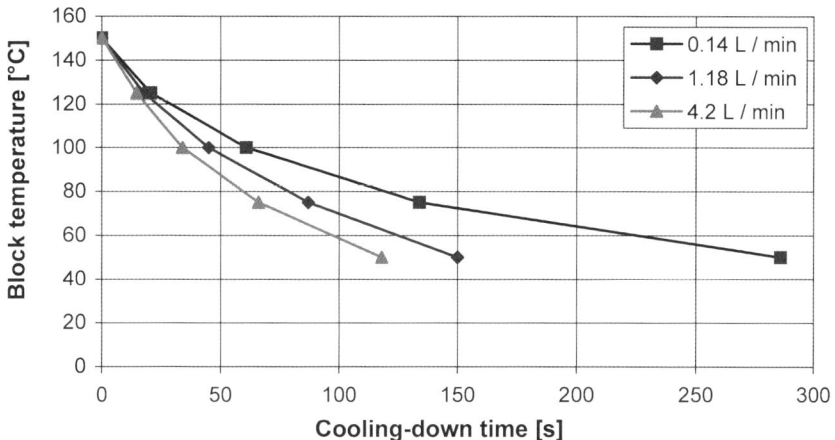

Figure 3.14 Dependence of cooling time on the flow rate of tap water.

these ovens so popular? Because they are rather simple, cheap and if the column has to be only kept at a constant temperature around ambient temperature, it is thought that the column temperature can be precisely controlled. Furthermore, the column can be installed very easily. It is also possible to include a column-switching valve in the oven. This is very convenient for automatic method development, because a large number of columns can be screened overnight to evaluate which column has the best separation performance for the analytical problem. A further advantage of these devices is the easy handling of the connections for the column within the oven. You only need to put the column in the oven and connect it with the inlet and outlet capillaries and that's it. However, the easy instalment bears the risk that the length of the connecting capillaries is not defined. It is possible that if the connecting capillaries have to be replaced, the length of the replacement capillary connecting the column to the injector may be different. This can lead to very subtle changes in the retention of target analytes, as discussed earlier, resulting in a complete failure of the method.

Although the air within the oven compartment can be heated up very rapidly, this is not true for the column because the heat transfer rate is much worse for air-bath heaters than for water jacket-heaters or block heaters.

Even if you only intend to use a column oven to work at ambient temperature, other problems may arise when using an air-bath oven. If the temperature of the laboratory cannot be accurately controlled and there are large temperature fluctuations, the precise control of the eluent and column temperature might be difficult. The reason is that the complete HPLC system cools down during the night, which means that the temperature of the solvents stored in the mobile phase reservoir can be 10 °C lower than during the day. Such a huge temperature change can lead to small but significant retention time shifts. We have observed this phenomenon in our own laboratory and many practitioners also confirmed that they had similar problems. The example which was presented in Chapter 2 (see Figure 2.1) might again serve as a good illustration. When fluorescence detection is used and the time after which the excitation and emission wavelength will be changed has been fixed, small shifts in the retention time of target compounds can mean that the analytes will not be detected even if they are present in the sample.

Another very serious problem arises when temperature programming is concerned. Although the air in the oven might be cooled down rapidly after a temperature programme, the temperature of the column will lag behind and hence, very long re-equilibration times result. Surely, one cannot speak of fast chromatography when it takes as long as 30 minutes or more until the next analysis can be started.

3.5.2 Water-Jacket Ovens

Some people think that fluid-based jacket ovens are very good for controlling the temperature of the mobile and stationary phases. The column as well as the preheating capillary can be easily placed in the fluid, which can be water or an oil of high boiling point, like silicone oil.[24] I think this works quite well if the column is operated from ambient temperature to slightly elevated temperature and water can be used as the heating fluid. However, if you plan to work at extremely high temperatures, the situation becomes more difficult. The advantage of using a water-jacket oven in contrast to an air-based oven is that the heat transfer is much better. Furthermore, the liquid circulates evenly around all the components including the capillary, the fittings and the column. This guarantees that there is a good heat transfer between the heating fluid and the column, as well as the preheating capillary. In principle all different column formats can be used and there is no need to worry about a special end fitting design. The question remains, however, as to whether the user is willing to place the column in a silicone oil bath. If there is a leakage due to a bad connection between the column and the capillaries, the heating fluid is contaminated and must be replaced. This is not only very time consuming but it may also result in your column being destroyed.

3.5.3 Block-Heating Ovens

As was described previously, in block-heating ovens the column and the capillaries are clamped between two aluminium blocks. These blocks have to be tailor made for each column, because a very tight contact has to be established in order to optimize heat transfer. Although this may be considered a big disadvantage, the additional effort in designing a high-temperature oven based on block heating will certainly pay off if temperature programming is applied. As long as the column is only operated under isothermal conditions, forced-air ovens might be used quite effectively if the mobile phase is adequately preheated. When temperature programming is used, however, a very efficient heat transfer has to be achieved in order to avoid the temperature of the mobile phase lagging behind that of the programmed gradient.

3.5.4 Summary

It is obvious that there are inherent advantages and drawbacks of the different technical approaches, both in terms of efficiency and applicability. The success of air-based column heaters is due to the fact that they are easy to operate and are relatively cheap. Air-based thermostats can also accommodate several columns and offer easy accessibility. The column only has to be placed inside the oven, which means that installation is very rapid. However, these systems are usually not equipped with a preheating device. This means that temperature changes in the laboratory can lead to a shift in the retention time. Moreover, temperature equilibrium is very slow when the oven temperature is changed. Although the air is heated up rapidly, the column temperature does not necessarily change as quickly. In order to get an impression of this effect, you should carefully observe the system pressure displayed by the pumps. In most cases you will notice that the pressure is not constant over a long period until a real temperature equilibrium inside the column is reached, although the oven temperature may have been constant for a while. So I would strongly advise you to closely monitor the system pressure. When the system pressure goes down even though the oven has reached the desired temperature, this is a clear sign that the column is not thermally equilibrated. Not until the pressure is constant can it be concluded that there is thermal equilibrium and that constant conditions have been reached. In case you are unsure, make two runs and overlay the chromatograms. If there are differences in the retention and you can exclude any effects related to malfunctioning pumps, the cause may be an insufficient thermal equilibrium between the column and the air in the oven, where an air-bath oven has been used. I would suggest that a third measurement is then made and compared against the last two chromatograms. If retention times are stable, then the system is thermally equilibrated. If there is another discrepancy you should analyse if there is a trend to shorter retention times, because in this case the system needs more time for thermal equilibration. As was demonstrated by de Villiers and Sandra, it might take as long as 60 minutes until the set temperature of 40 °C is reached when an Acquity column heater is used.[14]

A major drawback of fluid-based jacket column heaters is that the column has to be immersed into the liquid and free access to the column is restricted. In case of a leakage, this liquid will be contaminated with the mobile phase and has to be changed. In addition, the upper temperature limit is usually determined by the boiling point of the fluid. If water is chosen, the temperature range is limited to 100 °C. A higher column temperature can only be achieved if water is replaced by a silicone oil. However, this requires that the compartment into which the column is placed is sealed. Otherwise, the silicone oil may leak from the heating compartment. In my opinion, such a system is not well suited to industrial applications, although the heat transfer can be achieved very efficiently due to a very good circulation of the heating liquid.

A compromise has to be made when block heaters are used. In this case, heat transfer is most efficient and there is no problem of contamination of the heating liquid. However, most people think that by using an air-based column oven, the temperature can be changed very rapidly, but this is only true for the air inside the oven, not for the stationary phase itself. The heat transfer rate obtained with the metal-to-metal contact between the heating block and the elements being heated is far greater than can be obtained in an air-bath system, as long as the column fits the cavity exactly. This is very important when you plan to work in temperature-gradient mode, because the temperature of the stationary phase can be changed very rapidly. Nevertheless, the installation of the column and the connecting capillaries is not as easy as for the air and liquid column heaters and each column shape requires the fabrication of different shells. If all these aspects have to be weighed against each other, a block-heating device is the most suitable column oven in terms of heat transfer and fast temperature equilibrium. This also holds true if the temperature has to be decreased after a temperature gradient run.

References

1. S. Abbott, P. Achener, R. Simpson and F. Klink, *J. Chromatogr.*, 1981, **218**, 123.
2. J. D. Thompson, J. S. Brown and P. W. Carr, *Anal. Chem.*, 2001, **73**, 3340.
3. R. G. Wolcott, J. W. Dolan, L. R. Snyder, S. R. Bakalyar, M. A. Arnold and J. A. Nichols, *J. Chromatogr., A*, 2000, **869**, 211.
4. S. M. Fields, C. Q. Ye, D. D. Zhang, B. R. Branch, X. J. Zhang and N. Okafo, *J. Chromatogr., A*, 2001, **913**, 197.
5. T. Teutenberg, H. J. Goetze, J. Tuerk, J. Ploeger, T. K. Kiffmeyer, K. G. Schmidt, W. G. Kohorst, T. Rohe, H. D. Jansen and H. Weber, *J. Chromatogr., A*, 2006, **1114**, 89.
6. T. Teutenberg, W. gr. Kohorst and T. zu Höne, *Abschlussbericht Nr. 13522 N: Einsparung organischer Lösungsmittel und Effizienzsteigerung in der Flüssigchromatographie durch Einsatz der Hochtemperaturtechnik*, Arbeitsgemeinschaft industrieller Forschungsvereinigungen "Otto von Guericke" e.V., Köln, 2005.
7. D. Guillarme, S. Heinisch and J. L. Rocca, *J. Chromatogr., A*, 2004, **1052**, 39.
8. F. Gritti and G. Guiochon, *Anal. Chem.*, 2008, **80**, 5009.
9. A. Albrecht, R. Brüll, T. Macko, P. Sinha and H. Pasch, *Macromol. Chem. Phys.*, 2008, **209**, 1909.
10. L. C. Heinz, T. Macko, A. Williams, S. O'Donohue and H. Pasch, *LCGC Eur.-The Column*, 2006, **2**, 13.

11. H. Poppe, J. C. Kraak, J. F. K. Huber and J. H. M. van den Berg, *Chromatographia*, 1981, **14**, 515.
12. G. Mayr and T. Welsch, *J. Chromatogr., A*, 1999, **845**, 155.
13. G. Desmet, *J. Chromatogr., A*, 2006, **1116**, 89.
14. A. de Villiers, H. Lauer, R. Szucs, S. Goodall and P. Sandra, *J. Chromatogr., A*, 2006, **1113**, 84.
15. K. Kaczmarski, F. Gritti and G. Guiochon, *J. Chromatogr., A*, 2008, **1177**, 92.
16. F. Gritti, M. Martin and G. Guiochon, *Anal. Chem.*, 2009, **81**, 3365.
17. http://www.waters.com/waters/nav.htm?locale=en_US&cid=10002403 (last accessed October 2009).
18. http://www.chem.agilent.com/en-US/Products/Instruments/lc/systems/1290infinitylcsystem/Pages/default.aspx (last accessed October 2009).
19. http://las.perkinelmer.de/Catalog/CategoryPage.htm?CategoryID=Flexar+FX-15+UHPLC (last accessed October 2009).
20. http://www.thermo.com/com/cda/product/detail/1,,10123436,00.html (last accessed October 2009).
21. http://www.knauer.net/e/e_index.html (last accessed October 2009).
22. H. Poppe and J. C. Kraak, *J. Chromatogr.*, 1983, **282**, 399.
23. http://www.sim-gmbh.de/index.php?option=com_content&task=view&id=64&Itemid=502&lang=en (last accessed October 2009).
24. B. Yan, J. Zhao, J. S. Brown, J. Blackwell and P. W. Carr, *Anal. Chem.*, 2000, **72**, 1253.
25. T. Teutenberg, *Anal. Chim. Acta*, 2009, **643**, 1.

CHAPTER 4
Mobile Phase Considerations

In this chapter, I will discuss the effect of temperature on the mobile phase. Usually the practitioner does not have any information about the aggregate state of the mobile phase and its dependence on temperature and pressure. It is extremely difficult to search for this data in single publications. In many cases, this data are published in physical chemistry journals which are mostly overlooked by the chromatographic community. We have therefore collected and measured the most important data comprising the vapour pressure, the viscosity and the static permittivity over a large temperature interval for all the binary solvent mixtures, see Table 1.1. This data have been published in three consecutive papers in the *Journal of Chromatography, A*.[1-3] However, in this monograph I will mostly concentrate on the practical implications for the researcher at the bench. For a more in-depth study, I would advise the interested reader to consult the papers cited.

4.1 Influence of Temperature on Vapour Pressure

First of all I would like to repeat very briefly why it is necessary to acquire accurate data on the vapour pressure of the mobile phase. As I have already mentioned in the first chapter of this book, increasing the temperature of the mobile phase can lead to an unintentional phase transition of the mobile phase. To allow an accurate calculation of the temperature and pressure causing phase transitions in an HPLC system or in the separation column, knowledge of the vapour pressure of the mobile phases of pure solvents and binary mixtures is absolutely necessary.[i]

---

[i] The pressure of a gas at a defined temperature and in equilibrium with its condensed liquid phase is called "saturated vapour pressure" and is often described by presenting the vapour pressure as a function of temperature.

Mobile Phase Considerations

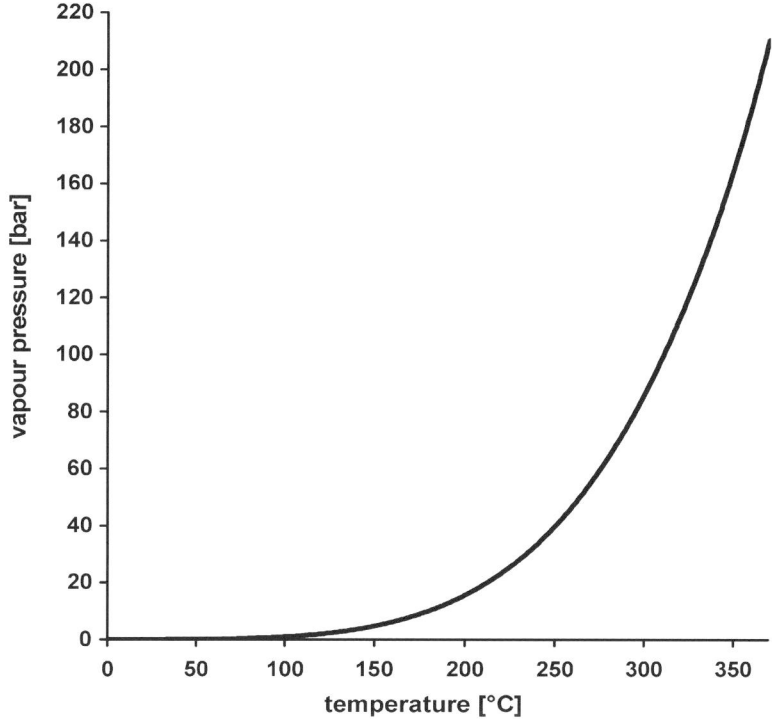

Figure 4.1 Dependence of vapour pressure on temperature for liquid water between 0 and 374 °C. (Calculated using parameters from ref. 4.)

In reversed-phase liquid chromatography, separations are not usually carried out with a mobile phase consisting of pure solvent.[ii] Instead, the mobile phase consists of water and an organic co-solvent. The fraction of the organic co-solvent is denoted by %B, which also highlights that this solvent is the stronger solvent. Water is considered to be the weakest solvent in reversed-phase separations and hence the elution strength of the mobile phase is varied by applying a solvent gradient. Usually the solvent gradient starts with a high water concentration of about 95% to 100% and ends with a high concentration of the organic solvent. Therefore a description of the bubble point line of binary eluent mixtures is needed, but how does the vapour pressure of these mixtures change at a defined temperature when the concentration of the organic solvent is increased? In most textbooks, the vapour pressure line of pure compounds can be easily found. Figure 4.1 depicts the vapour pressure of water as the temperature is increased from 0 °C up to the critical temperature at 374 °C.[4]

[ii] Exceptions will be given in Chapter 8 where I present important hyphenation techniques which are based on the use of a pure water mobile phase. However, for conventional reversed-phase separations at high temperatures, the use of an organic co-solvent is highly recommended.

It is obvious that the vapour pressure does not increase linearly but exponentially as the temperature is raised. For water, the vapour pressure is around 1 bar when the temperature is not higher than 100 °C. Therefore, you do not have to worry much about a phase transition of water within the column. However, bubble formation might already take place in the detector cell if the eluent is not cooled down as it enters the detector. Hence a back-pressure regulator should be installed behind the cell in order to circumvent this problem as was already mentioned in Chapter 2.

Before I explain the procedure for preventing a phase transition in detail, by highlighting the necessity of back-pressure control, I will give a helpful example of an incident that happened in our laboratory. One of my students was working on a project on high-temperature HPLC. He was very experienced but one day made a small mistake. He turned on the oven and adjusted the temperature to 185 °C without activating the pumps. Since he used a block-heating oven, the set temperature was reached very rapidly. It was only after a period that he realized that he had forgotten to turn on the pumps. This means that while thermal equilibrium between the heating block and the column was already established, there was no flow through the system. So in effect the column was boiled. He then activated the column cooling and started the flow, but it was too late. The effluent had a yellow colour and after a test sample was run the fate of the column was revealed.[iii] The column my student used in that experiment was a polybutadiene-coated zirconium dioxide column. But the same phenomenon was observed when a bare zirconium dioxide column, a normal-phase column without any surface modification, was used. This time it was not the fault of the operator but a pump failure which led to the degradation of the column. During a column test study where the temperature was adjusted to 185 °C, the pumps were switched off erroneously and the flow stopped while the oven temperature was maintained at 185 °C. Although the problem was noticed after a few minutes, it was again too late when the flow was restarted. The effluent was yellowish and the separation of a test mixture revealed that the column could no longer be used. These two examples serve to highlight that a precise back-pressure control is really of utmost importance. It becomes also clear that the oven should be integrated into the software of the HPLC system so that an automatic shut-down and cooling of the column starts as soon as there is a problem with the flow. At the time we performed these experiments, there was no communication between the pump and the heating oven.

So my first instruction is that you should always switch on the pumps so that a flow and also a pressure are generated. For a better understanding of the pressure in an HPLC system, the facts summarized above are illustrated in Figure 4.2.

[iii] I think that this is the most efficient way to get rid of columns you don't like. If your boss tells you to run an analysis on a column you don't consider to be appropriate, but you can't convince him that he's not right, just put this column in an oven, heat it up so that the mobile phase boils inside the column and then start the flow. I guess that most materials cannot be used after this treatment. Even very rugged materials like metal oxide columns are destroyed by this procedure.

Mobile Phase Considerations

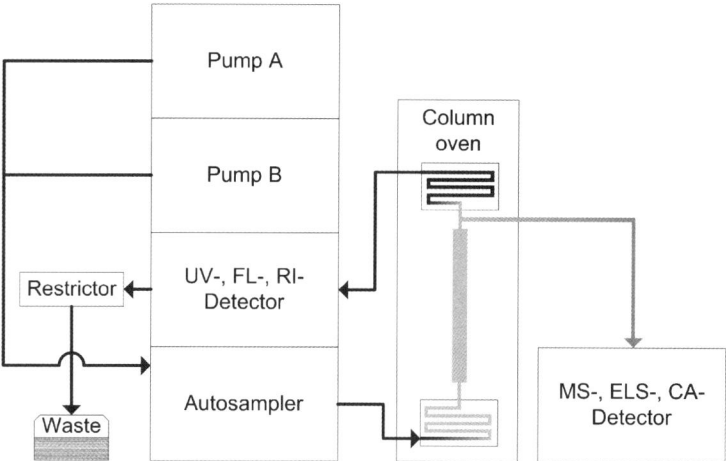

Figure 4.2 Scheme of a typical HPLC system with a modular heating oven for high-temperature operation.[1] (Reproduced with kind permission from Elsevier.)

Here, the scheme of a typical HPLC system is given. Usually, the eluent is mixed using two pumps and a mixing chamber. Afterwards, the eluent is introduced to the autosampler. In case of high-temperature liquid chromatography, preheating of the mixing chamber and the autosampler is not necessary as was already outlined in Chapter 3. A short preheating capillary is sufficient to match the temperature of the mobile phase to the temperature of the stationary phase when a suitable heating device is used. This heating system incorporates the concept of a modular oven I have outlined in the previous chapter. The eluent then runs through the column, which is kept at the desired temperature. What has to be considered is that although the inlet pressure can be 1000 bar or even higher, a pressure gradient forms along axial direction of the column dropping to atmospheric pressure at the end of the column if there are no further connecting capillaries. Therefore, a phase transition of the eluent can occur at the point in the column where the vapour pressure of the eluent is higher than the pressure generated by the remaining particles of the packed bed. Usually the column is connected with a detector using a transfer capillary, which generates a small back pressure. The length of this capillary depends on the system set-up and should be as short as possible to reduce the system volume and hence peak broadening. The pressure which is generated by this capillary not only depends on the length and internal diameter of the capillary. Other factors, such as the flow rate, the mobile phase, the temperature and even the surface roughness of the capillary wall, also contribute to the pressure drop across the tubing. This means that an exact calculation of the local pressure drop within the capillary is quite complex. The easiest way to verify if there might be a phase transition in the capillary would be to disconnect the column and only measure the pressure which is generated across the capillary.

But consider that even the temperature profile in the axial direction of this capillary must be known. If you would like to work in temperature gradient mode, the temperature of the column is increased during the chromatographic run. This means that the temperature profile in the axial direction of the transfer capillary also changes with time. The reason is that at the start of the temperature gradient the eluent exits the column with a low temperature, while at the end of the temperature gradient the eluent exits with a high temperature. If we do gas chromatography and apply a temperature gradient, we don't have to worry about the change in temperature during the run because the mobile phase is already a gas and thus no phase transition can occur. But in liquid chromatography, this is a little bit different. So how can we overcome this problem? The easiest way is to use a heating system where the eluent is cooled to ambient temperature after it has left the column. Fortunately, some column ovens offer the possibility to cool down the eluent before it enters the detector. This is necessary when a detector is used which is very sensitive to small fluctuations of the eluent temperature, like a refractive index or fluorescence detector. An example has already been given in the previous chapter, where the effect of eluent temperature on the peak area of test compounds was highlighted for UV detection (see Figure 3.12). Eluent cooling prior to detection can then be a prerequisite to enhance the sensitivity of the detection process. This means that the temperature of the eluent is around 30 °C when the mobile phase enters the detector. In this case, a phase transition might only be possible in the short distance from where the hot eluent leaves the column until it enters the post-column cooling unit of the heating system. In order to be absolutely sure that even there a phase transition cannot occur, a back-pressure regulator should be installed.

It can be summarized that by using a heating system with a post-column eluent cooling and a back-pressure regulator which is placed behind this cooling unit, you can be absolutely sure that there is no phase transition of the mobile phase in your system.

However, if you would like to optimize your system in terms of reducing the extra column volume, a back-pressure regulator might not always be the best choice. In the following sections the implications of the hyphenation of a high-temperature liquid chromatographic system with different HPLC detectors, in terms of a back-pressure control, are discussed. The two possibilities to maintain a sufficiently high outlet pressure by either using a back-pressure regulator or a restriction capillary will be considered.

4.1.1 Prevention of a Phase Transition using a Back-Pressure Regulator

The UV detector is still the most widely used detector in HPLC separations, despite mass spectrometry becoming increasingly important. If UV detection is employed, a back-pressure regulator may be installed behind the detection cell since nearly all instruments are equipped with cells which can tolerate up to

Mobile Phase Considerations

40 bar. Cells designed for supercritical fluid chromatography have a pressure limit of 300 bar or even higher. However, the technical specifications of the manual should always be thoroughly studied before a back-pressure regulator is installed behind the detector. Otherwise, irreversible damage to the cell might occur. If the mobile phase is cooled down to ambient temperature prior to detection, referred to as post-column cooling, a back-pressure regulator placed behind the detection cell will keep the outlet pressure above the vapour pressure of the mobile phase.

Now, let's look what happens if you run a solvent gradient using a binary mobile phase at high temperature. I will concentrate here on the two solvent systems, water–acetonitrile and water–methanol, because these systems are by far the most popular mobile phases when reversed-phase HPLC is concerned. Further data for additional hydro-organic mixtures is compiled in Appendix A.

Figure 4.3 shows the vapour pressure line for the water–acetonitrile system, while Figure 4.4 depicts the vapour pressure line for the water–methanol system.

What happens to the vapour pressure when a linear solvent gradient is applied? For the water–acetonitrile system, the vapour pressure runs through a maximum and then decreases again. It is interesting to note that for lower

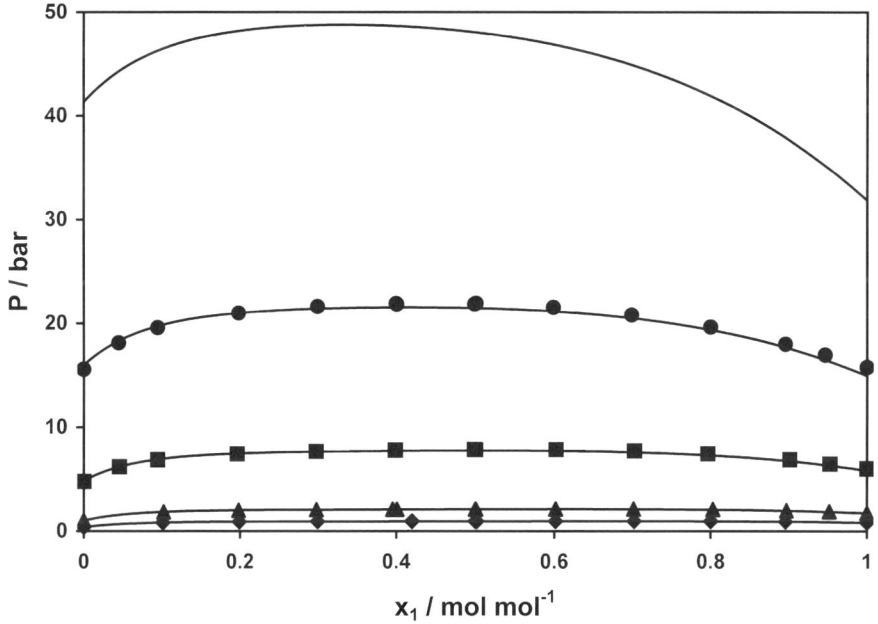

Figure 4.3 Experimental isothermal P-x-data of acetonitrile (1)–water (2) at different temperatures: ◆, 75 °C; ▲, 100 °C; ■, 150 °C; ●, 200 °C; and 250 °C, correlated with NRTL (—). The pressures for 250 °C have been calculated with the constants given in ref. 1. (Reproduced with kind permission from Elsevier.)

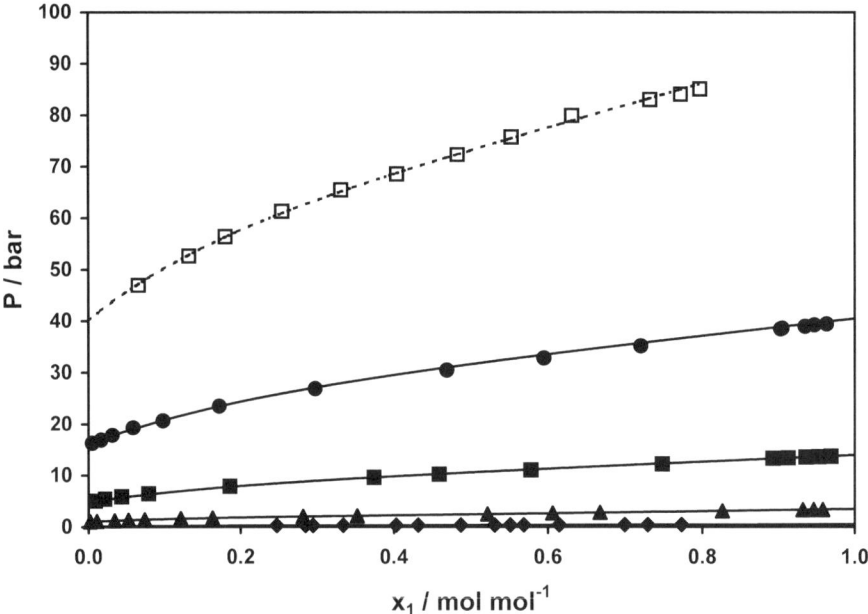

Figure 4.4 Published isothermal P-x-data of methanol (1)–water (2) at different temperatures: ◆, 50 °C; ▲, 100 °C; ■, 150 °C; ●, 200 °C; and □, 250 °C, correlated with NRTL (—) or PSRK (--).[1] Literature data have been taken from DDB 2008. (Reproduced with kind permission from Elsevier.)

temperatures, there is only a marginal increase in the vapour pressure as the concentration of acetonitrile is raised. In contrast to this, the absolute increase in the vapour pressure is more pronounced the higher the temperature. Therefore, working at very high temperatures above 100 °C requires the knowledge of the change of the vapour pressure of the binary mixture with temperature, because it can no longer be assumed that the vapour pressure remains constant during the solvent gradient. When methanol is used instead of acetonitrile as the organic co-solvent, it is striking that the vapour pressure continuously increases the higher the concentration of methanol. This means that the lowest vapour pressure is always noticed for pure water, whereas the highest vapour pressure is observed for pure methanol.

In the following paragraph, I will explain in detail what happens if a solvent gradient is applied consisting either of water–acetonitrile or water–methanol at a temperature of 200 °C. The conclusions which can be drawn from the discussion of these examples can then be easily transferred to other binary solvent systems compiled in Appendix A. If we start with pure water, the back pressure has to be above 16 bar at 200 °C. By gradually changing the composition of the mobile phase to 100% acetonitrile, the vapour pressure reaches a maximum of about 22 bar at 42 mol% of acetonitrile and then decreases to 15 bar as pure acetonitrile is approached. In order to guarantee that there is no phase transition during the solvent gradient, the back pressure should always be higher than the highest vapour pressure of the mixture. In the case where acetonitrile is used as the organic solvent, this is usually at an intermediate solvent composition. If you use methanol instead of acetonitrile, the highest vapour pressure occurs when pure methanol is used. For a mobile phase consisting of water and methanol at a temperature of 200 °C, the back pressure should therefore be higher than 40 bar. *I strongly advise the practitioner to adjust the back pressure to the highest vapour pressure of any composition of the solvent mixture at a given temperature. This guarantees that even in the case of a pump failure, where the composition of the mobile phase is changed erroneously, a phase transition in the system will not occur immediately. Please make sure that the high-temperature oven is integrated in the software of the HPLC system, so that a pump failure will also trigger a shut down of the oven. Otherwise, the stationary phase will be irreversibly damaged.*

If the back pressure is always higher than the vapour pressure of the mobile phase at a given column temperature, there is no need to worry about a phase transition. Also, the pressure drop across the column is irrelevant as long as the back pressure applied at the column or detector outlet is above the vapour pressure of the mobile phase. According to the data given in Figures 4.3 and 4.4, at 200 °C the highest pressure occurs if pure methanol is used as a mobile phase and is about 40 bar. Certainly pure methanol is not a typical mobile phase in reversed-phase HPLC, but you have to consider that most separations are carried out in solvent-gradient mode. Hence the composition of the mobile phase is continuously changed during the chromatographic run, and in many cases you end with 100% of the organic modifier (100% B) in order to elute non-polar compounds.

In cases where the temperature is increased further, the exponential rise of the vapour pressure with temperature can cause problems. For the water–tetrahydrofuran system we have experimentally determined the vapour pressure at temperatures up to 250 °C. In Figure 4.5, the exponential increase is evident, resulting in a maximum pressure of about 62 bar for a mobile phase composition of approximately 50 : 50 mol%. However, for UV detection this pressure should not present a problem if a suitable pressure-resistant flow cell is used.

Unlike UV detection, the cells of fluorescence detectors are in most cases able to withstand only moderate pressures of about 5 to 20 bar. This means that when working at eluent temperatures above 150 °C, the vapour pressure of the eluent mixture is above the tolerable pressure of the cell. In this case, the back-pressure regulator has to be installed before the detector cell. The disadvantage of connecting the restrictor before the detector is that it contributes to the extra-column volume and compromises the chromatographic resolution. In order to circumvent this problem, a closer look on the data presented in Figures 4.3 and 4.4 will help to find the best operating conditions.

Let us assume that the column is operated at 150 °C. The vapour pressure of pure water at 150 °C is approximately 5 bar, whereas the highest vapour pressure at this temperature is encountered for pure methanol and is about

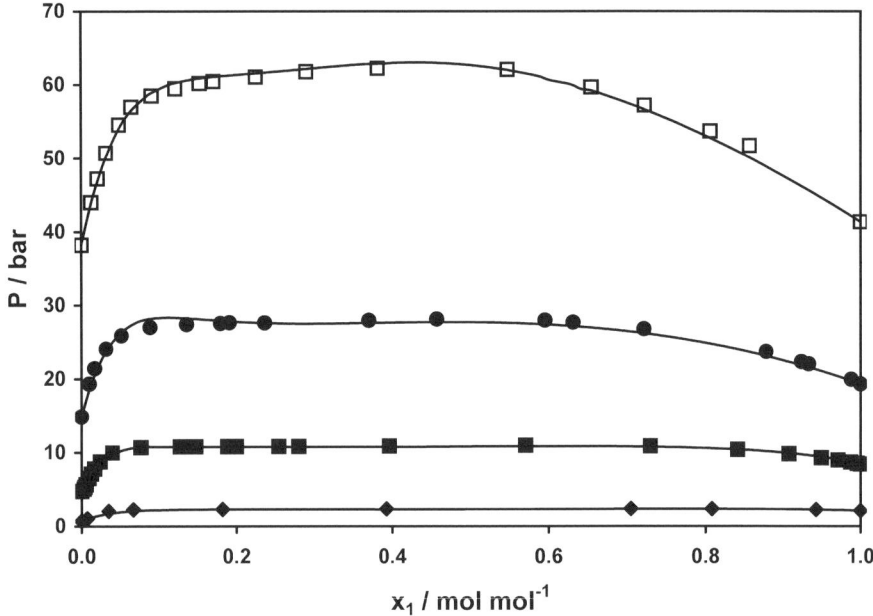

Figure 4.5 Experimental isothermal P-x-data of tetrahydrofuran (1)–water (2) at different temperatures: ◆, 90 °C; ■, 150 °C; ●, 200 °C; and □, 250 °C, correlated with NRTL (—).[1] (Reproduced with kind permission from Elsevier.)

14 bar. If it is planned that a solvent-gradient elution will be performed using a binary mobile phase of water and methanol, the restrictor should at least be adjusted to 15 bar. In the case where the HPLC system is only run with water and acetonitrile, the maximum vapour pressure will be below 7 bar, which means that it would be possible to place the restrictor behind the cell. *However, it must be explicitly pointed out that the technical manual has to be consulted. The connection of the restrictor behind the fluorescence detector also has to be done with great care; otherwise the cell will be irreversibly damaged.* The need to place the restrictor behind the cell clearly depends on the requirements of the method. If the resolution of a critical peak pair has to be optimized, the reduction in the dead volume might be essential, which means that the restrictor should be placed behind the cell. If resolution is not a problem, the restrictor might be placed before the cell. The data presented in Figures 4.3 to 4.5 enable the user to decide whether to place the restrictor before or behind the detector.

However, the use of a back-pressure regulator is not the only option for back-pressure control. The back pressure can also be adjusted by using a transfer or connecting capillary with a very small internal diameter. This strategy is discussed in the next section and is highly recommended for evaporative light-scattering, charged-aerosol and mass spectrometric detection.

4.1.2 Prevention of a Phase Transition using a Restriction Capillary

Evaporative light-scattering detection is usually used for analytes lacking a chromophoric system where UV detection is not feasible. The mobile phase is converted into a gas and analyte particles are created, which are then detected by light scattering. It must be noted that the gas flow and the temperature of the nebuliser gas have to be adapted to the organic co-solvent in the mobile phase. Hence, solvent-gradient elution often exhibits a problem, because it has been reported that the nebulisation and the sensitivity of the detection process is significantly influenced by the composition of the mobile phase.[5,6] Thus, heating the mobile phase close to its transition temperature from a liquid to the gas phase might be advantageous in terms of detection sensitivity, as was also suggested by Guillarme *et al.*[7] This also means that it will not be necessary to cool down the eluent after it leaves the column as long as the phase transition is avoided in the transfer capillary connecting the column with the detector. However, the pressure which is generated by this capillary also strongly depends on the temperature of the mobile phase. For practical reasons, the pressure which is generated by this capillary should be measured without installing the column. A complete theoretical treatment of all the phenomena related to a precise description of the course of the temperature of the mobile phase in the capillary, and its dependence on the length and inner diameter of the capillary as well as the flow rate and temperature of the mobile phase, is far too complex and beyond the scope of this monograph. It should be made clear that the pressure at the end of the capillary should be higher than the vapour pressure of the mobile phase entering the detector. A simple experiment should be carried out where a test solute is injected and then detected without a column. The connecting capillaries should be placed in the column oven, and the column should be replaced by a zero dead-volume connection. The peak shape of the test solute should then be monitored during the course of the gradient. If the peak shape has not deteriorated, which means that split or distorted peaks are not detected, the back pressure created by the transfer capillary is sufficient to prevent a phase transition of the mobile phase.

In order to better understand this effect, a simple experiment has been performed to visualize this phenomenon. Figures 4.6 and 4.7 show two chromatograms which were obtained using a time-of-flight mass spectrometer (TOF-MS) hyphenated to a high-temperature HPLC system. A test sample containing several pharmaceutical compounds has been eluted with a mobile phase consisting of water and acetonitrile in solvent-gradient mode at 90 °C (Figure 4.6) and 120 °C (Figure 4.7), respectively. The transfer capillary connecting the column with the inlet source of the mass spectrometer was exposed to room temperature and consisted of a stainless-steel tube with an inner diameter of 127 µm and a length of 35 cm. As the temperature was increased to 120 °C, the peaks appear to be distorted and split, which is clearly visible in Figure 4.7. A general assumption was that the column has been degraded at that temperature. However, a further run at 90 °C could not verify this

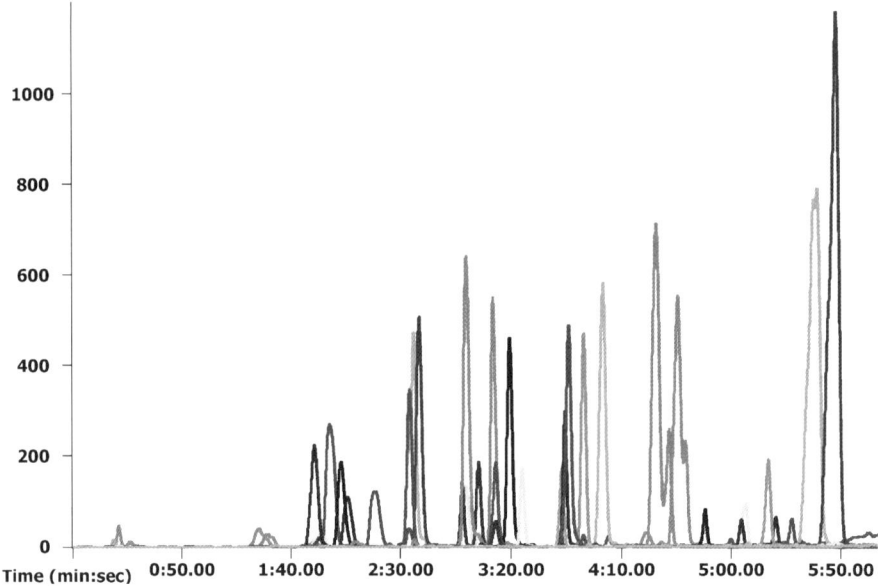

Figure 4.6 Separation of a test mixture containing pharmaceutical compounds. Detection: LECO Unique time-of-flight mass spectrometer. Chromatographic conditions: flow rate: $1\,\text{ml}\,\text{min}^{-1}$; solvent A: water (containing 0.1% formic acid); solvent B: acetonitrile (containing 0.1% formic acid); solvent gradient: 0 to 100% solvent B in 10 min; temp.: 90 °C; column: Zorbax StableBond (5 cm × 3.0 mm ID; 1.8 µm); transfer capillary: 127 µm ID, length of 35 cm.[1]

assumption. Therefore, it could be shown that bubble formation and hence a phase transition of the mobile phase within the transfer capillary was responsible for this phenomenon.

These findings are also relevant for the hyphenation of a high-temperature HPLC system with a flame-ionization detector. Similar results were reported by Ingelse and co-workers, who clearly confirmed the effects which I have displayed in Figures 4.6 and 4.7. The authors noted that even if the column was operated isothermally, the use of linear restrictors occasionally resulted in detector spiking due to evaporation and subsequent condensation of the eluent in the restrictor, which they termed "sputtering".[8] This became even worse when temperature programming was applied. The use of tapered restrictors, as frequently used in supercritical fluid chromatography, helped to eliminate the problem of solvent sputtering but suffered from an easy blockage. Therefore, they only used linear restrictors but placed the transfer capillary in a second oven which was kept at a constant temperature below the boiling point of water. This example again clearly demonstrates that the system needs to be carefully designed. The question as to whether the restrictor has to be placed outside the column oven also depends on the method. Provided that the temperature is not increased above the normal boiling point of the mobile phase,

Mobile Phase Considerations

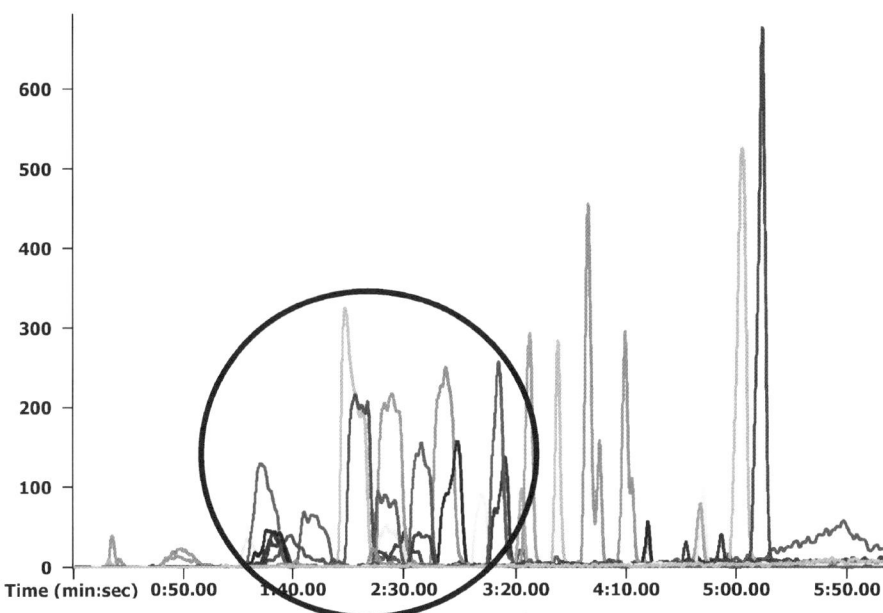

Figure 4.7 Separation of a test mixture containing pharmaceutical compounds. Detection: LECO Unique time-of-flight mass spectrometer. Chromatographic conditions: flow rate: $1\,\text{ml}\,\text{min}^{-1}$; solvent A: water (containing 0.1% formic acid); solvent B: acetonitrile (containing 0.1% formic acid); solvent gradient: 0 to 100% solvent B in 10 min; temp.: 120 °C; column: Zorbax StableBond (5 cm × 3.0 mm ID; 1.8 μm); transfer capillary: 127 μm ID, length of 35 cm.[1]

there is no need to worry about a phase transition, a solvent perturbation or a sputtering effect. However, if a large temperature interval is considered and the temperature is increased up to 200 °C or even higher, a set-up as described by Ingelse seems to be very well suited to avoid such problems.

If temperature programming is used, it should be noted that starting with a low temperature might lead to an excessive back pressure if a capillary with a very small internal diameter is used as a linear restrictor. Hence, the approach described by Ingelse to use a transfer line which is heated independently from the column will guarantee that neither an excessive back pressure is observed at a low column temperature, nor that there is a problem related to a phase transition of the mobile phase at a high column temperature. With this set-up, a constant back pressure is obtained from the restrictor regardless of the temperature programme.

It can be summarized that there are two alternatives to prevent the mobile phase from boiling when working at very high temperatures. The first one is to use a back-pressure regulator. If possible, the back-pressure regulator should be mounted behind the detector in order to minimize the extra-column band broadening. The second option is to use a restriction capillary which acts as a linear restrictor. *The requirement of a back-pressure control strongly depends on the detector and the heating system.* These examples clarify that besides the knowledge of the vapour pressure, data on the temperature dependence of the viscosity is also needed. The results of these measurements and the implications for high-temperature HPLC are described in the next section.

4.2 Influence of Temperature on Viscosity

Viscosity is a measure of the flow behaviour of a liquid. The higher is the viscosity, the less volatile is the liquid. With increasing temperature, the viscosity strongly decreases because intermolecular interactions like van-der-Waals interactions or hydrogen bonds will break due to the higher kinetic energy of the molecules.

In liquid chromatography, the viscosity of the mobile phase is generally larger than in gas chromatography. This is a big drawback because the high viscosity causes low diffusion coefficients and strong flow resistances of the mobile phase resulting in a high back pressure and a high mass transfer resistance.

For a pure compound, *e.g.* liquid water, the viscosity decreases with increasing temperature. This is highlighted in Figure 4.8, where the viscosity of pure water is depicted as the temperature is increased from 0 °C to 374 °C.[4]

The decrease in viscosity is not linear but exponential. This means that up to approximately 100 °C, a very strong decrease is noticed, while at temperatures above 100 °C the decrease is much flatter. As I have stressed before, in conventional reversed-phase HPLC, a binary mobile phase consisting of water and an organic co-solvent is typically used. When a solvent gradient is applied at ambient temperature, a more or less steep increase in the system pressure is observed. This increase in the system pressure strongly depends on the organic

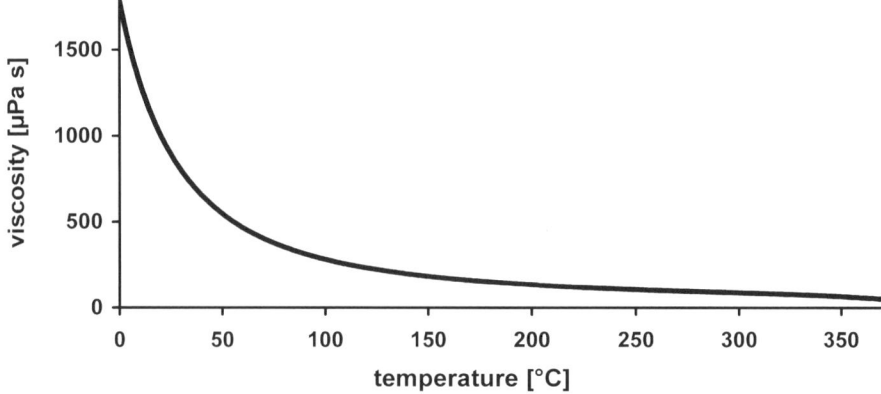

Figure 4.8 Dependence of viscosity on temperature for liquid water between 0 and 374 °C. (Calculated with parameters from ref. 4.)

co-solvent. Every practitioner probably knows that a binary mixture of water–methanol is much more troublesome than the respective mixture of water–acetonitrile in solvent-gradient mode. The reason is that most liquid chromatographic methods are run at ambient temperature or slightly elevated temperatures. In this region, the water–methanol system exhibits a huge viscosity maximum as the concentration of the organic solvent is changed during the solvent gradient. I guess that in many cases, this has led to a widespread reluctance to use methanol instead of acetonitrile. Fortunately, as the temperature is increased, the huge "excess" viscosity which is experienced at ambient temperature is much lower at elevated temperatures. In a recent research project, we have acquired data on the viscosity for all the solvent systems I have listed in Table 1.1 in a broad temperature range.[9] As for the back pressure, I will not go into detail about the experiments to generate the data, but will instead focus on some general remarks about the change of the viscosity with temperature. Again, I will concentrate on the two most commonly used solvent systems, water–acetonitrile and water–methanol. For the other solvent systems listed in Table 1.1, you will find more information in Appendix B.

When comparing the data which is presented in Figures 4.9 and 4.10, it can be clearly seen that a mixture of water–acetonitrile is much less viscous than a mixture of water–methanol at ambient temperature. While the viscosity maximum for water–acetonitrile at 25 °C is around 1000 µPa · s, the corresponding viscosity maximum for a mixture of water–methanol is around 1600 µPa · s. This means that the pressure will be much higher if a solvent gradient of water–methanol is run instead of water–acetonitrile at otherwise

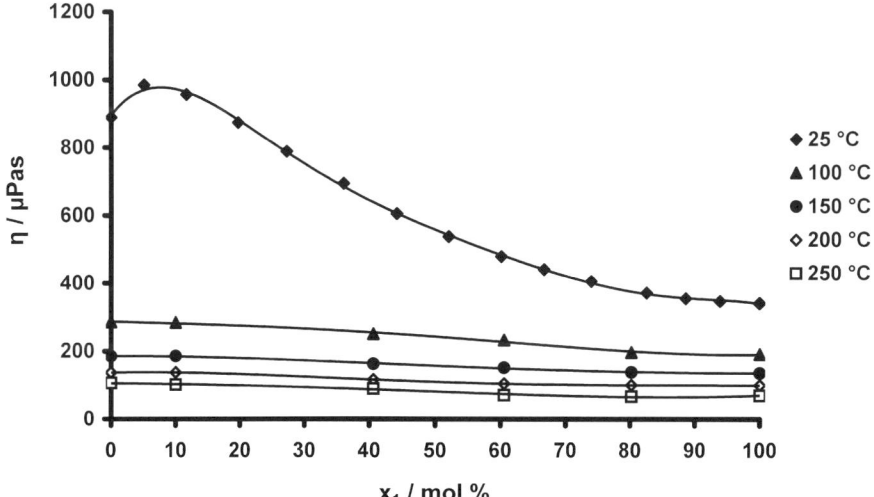

Figure 4.9 Experimentally determined viscosities of the binary mixture acetonitrile (1)–water (2) at different temperatures and 100 bar.[2] (Data at 25 °C have been taken from literature.)

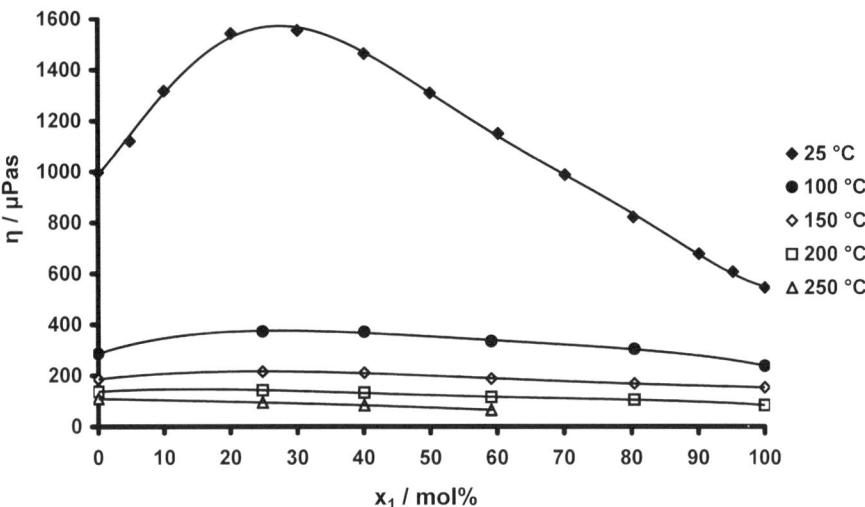

Figure 4.10 Experimentally determined viscosities of the binary mixture methanol (1)–water (2) at different temperatures and 100 bar.[2] (Data at 25 °C have been taken from literature.)

constant conditions. However, as the temperature is increased, the huge viscosity maximum becomes smaller. At 100 °C, there is only a minimal increase in the viscosity for the mixture water–acetonitrile. In fact, the viscosity only changes between 200 and 300 µPa · s. The same trend holds true for the mixture of water–methanol at 100 °C. Here, the viscosity changes between 300 and 400 µPa · s. When compared with data at ambient temperature, it can be seen that the viscosity at 100 °C is reduced by more than three-fold for the water–acetonitrile system and by more than four-fold for the water–methanol system. Interestingly, if both solvent systems are compared at temperatures above 150 °C, virtually no "excess" viscosity is observed. Also, both solvent systems exhibit nearly the same viscosity. What is the practical impact of these results? For method development, you don't have to worry about a viscosity maximum which will be experienced at ambient temperature. Since there is hardly a difference between water–acetonitrile and water–methanol mixtures at temperatures above 150 °C, both mixtures can be used under the same conditions. This means that a constant flow rate generates the same maximum pressure for both mixtures.

Without going into further detail, I would only like to mention that the same rules apply for the water–ethanol system, which exhibits a viscosity maximum around 2400 µPa · s at 25 °C. At 100 °C, the viscosity maximum is around 500 µPa · s and at 250 °C it is around 100 µPa · s. Although this solvent mixture might be regarded as very exotic by most practitioners, I will demonstrate in Chapter 8 that there is a new hyphenation technique which is known as LC taste[R] and relies on this specific solvent system. *The data on the viscosity of the different mobile phases I have given in Appendix B reveals that all solvent systems* can be used at elevated temperature in solvent-gradient mode. The huge "excess" viscosity which is very pronounced for the alcoholic modifiers is completely reduced at high temperatures.

I would now come back and pick up the problem I discussed earlier in section 4.1, where a solvent perturbation was noticed which was caused by a phase transition of the mobile phase in the transfer capillary connecting the HPLC system to the mass spectrometer.

4.2.1 Practical Implications – The Restriction Capillary

As I have pointed out in section 4.1, a simple restriction capillary can be used to control the back pressure at elevated temperatures when an HPLC system is hyphenated to a mass spectrometer. However, the calculation of the pressure drop across a capillary requires the knowledge of the viscosity as is expressed by the following equation, which is also known as the Hagen–Poiseuille equation:

$$\Delta P = \frac{128 \cdot \eta \cdot l \cdot F}{d_i^4 \cdot \pi} \quad (4.1)$$

Here, η is the viscosity of the mobile phase; l is the length of the capillary; F is the flow rate of the mobile phase; and d_i is the inner diameter of the capillary. Since the diameter of the capillary in the equation changes with the 4th power, the diameter can be adjusted to the viscosity of the mobile phase at a given temperature to prevent the mobile phase from boiling. A problem which is not often properly addressed arises when the temperature and the composition of the mobile phase are not constant as a result of solvent and temperature programming. In both cases, the pressure in the capillary will change. This can cause solvent perturbations within the capillary due to bubble formation, leading to distorted peak shapes as was shown in Figure 4.7. In order to suppress such a phase transition, you have to use a capillary with a smaller internal diameter. For the system set-up described in Figure 4.7, a transfer capillary with an internal diameter of 127 μm and a length of 35 cm was used. The temperature of the mobile phase was 120 °C as it left the column. Because the transfer capillary connecting the column with the inlet source of the mass spectrometer was exposed to room temperature, it can be assumed that there is a strong cooling effect and the eluent will not reach the detector with such a high temperature.[iv] In this example, a mobile phase of water and acetonitrile

---

[iv] I would like to refer the reader to Chapter 3 where I have demonstrated that the mobile phase cools down rapidly if the capillary is exposed to room temperature after it leaves the heated zone of the oven. This means that there is a viscosity gradient in the axial direction of the capillary. Hence, if you would like to exactly calculate the pressure drop across the length of the capillary, the viscosity gradient should be taken into account. This means that the real pressure drop will be higher in the case where the capillary is exposed to room temperature than where it is heated to a constant temperature. Please note that the heating of a transfer line is very important for GPC analyses, because the solubility of polymers strongly depends on temperature. In contrast to this, solubility problems do not usually occur in reversed-phase liquid chromatography once the sample is dissolved in a suitable solvent.

was used at 120 °C, which means that the highest vapour pressure is 3.2 bar at 60 mol%. Also, solvent programming was employed, running a gradient from 0 to 100% acetonitrile over 10 minutes at a flow rate of 1 ml min^{-1}. At this temperature, the lowest viscosity of the mobile phase is 242 µPa · s when the concentration of acetonitrile is raised from 0 to 10 mol%. In order to avoid a phase transition, the lowest viscosity has to be taken into account. By including all these parameters in Equation (4.1) and keeping the flow rate constant, a pressure drop of about 2.2 bar results. Clearly, this is not sufficient to prevent the boiling and phase transition of the mobile phase. Therefore, the transfer capillary was replaced and a 63 µm ID capillary was used instead, yielding a total pressure drop across the capillary of approximately 37 bar. Then, the experiment described in the previous section was repeated again. The resulting chromatogram is displayed in Figure 4.11. It can be clearly seen that all peaks now elute symmetrically without any evidence that the column has been degraded. Therefore, it was not the column which produced the distorted peaks, but a solvent perturbation within the transfer capillary has led to a partial phase transition of the mobile phase.

The problems described here may sound trivial to the experienced practitioner. However, if you are not familiar with high-temperature HPLC, it is not

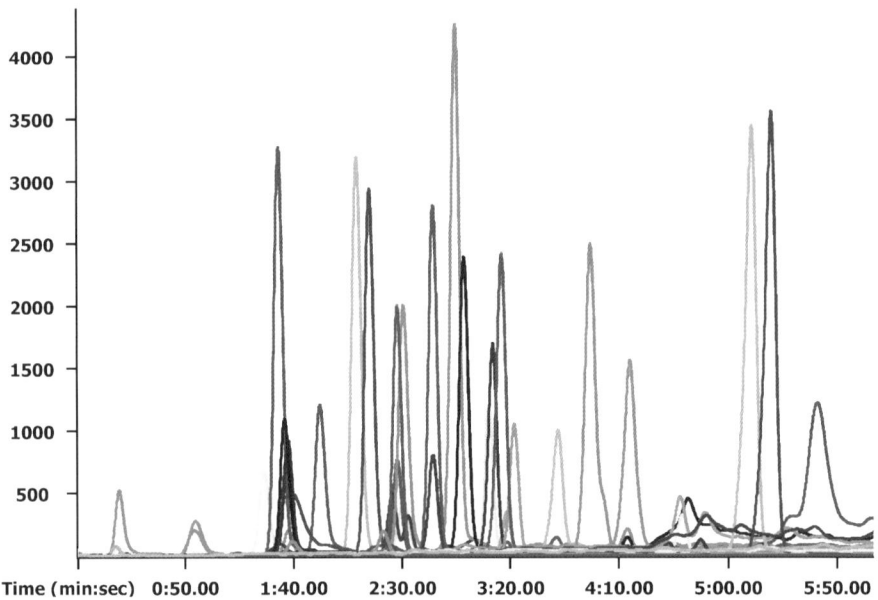

Figure 4.11 Separation of a test mixture containing pharmaceutical compounds. Detection: LECO Unique time-of-flight mass spectrometer. Chromatographic conditions: flow rate: 1 ml min^{-1}; solvent A: water (containing 0.1% formic acid); solvent B: acetonitrile (containing 0.1% formic acid); solvent gradient: 0 to 100% solvent B in 10 min; temp.: 120 °C; column: Zorbax StableBond (5 cm × 3.0 mm ID; 1.8 µm); transfer capillary: 63 µm ID, length of 35 cm.[2]

easy to distinguish between effects which are caused by the column and problems which can be related to extra-column effects. When looking at the chromatogram depicted in Figure 4.7, the first guess probably will be that the column is destroyed by the high temperature. Therefore, if you are not sure if the column is the problem or any other part of the HPLC system, run a test mixture at defined conditions and compare the efficiency of the column. I would always suggest the use of a control chart in order to evaluate and monitor the performance of the column. A simple test mixture is sufficient, but you can also run more sophisticated tests like the Engelhardt or Neue test or procedures according to Sander.[10–12]

While in this paragraph the focus was on the implications of temperature on the hardware of the HPLC system, the question remains as to how temperature affects the speed of a separation? Since this question is not as trivial to answer as it might appear, the most important facts will be discussed in the next section.

4.2.2 Practical Implications – Kinetic Aspects and Column Pressure

Although this is a monograph about high-temperature HPLC, the rapid technological development of the last five years make it necessary to consider both the variables of temperature and pressure in order to discuss the data I have presented in the last section. Indeed, the introduction of HPLC systems which are capable of delivering a maximum pressure of up to 1200 bar at high flow rates up to $5\,\mathrm{ml\,min^{-1}}$ must be warmly welcomed. For many analyses where only a few analytes have to be separated and selectivity is not a big problem, methods can be enormously speeded up by such systems leading to savings in both analysis time and solvent consumption. A discussion about the kinetic benefit of high eluent temperatures is also linked with the possibility of running a method at a high pressure. At this point, I would like to emphasize that temperature and pressure are not opposed, but indeed can be used in a complementary fashion. Often, separation scientists like to categorize methods and technologies which in effect rely on the same principles. Here, a unified approach would be more helpful.[v,13]

---

[v] A very nice concept has been presented by Chester, which is entitled "Unified Chromatography". Chester presented this concept for multidimensional chromatographic separations. However, the advantage of this approach is that every technique, regardless of whether it is termed HPLC, UHPLC, HTLC, GC or SFC can be described by using the same phase diagram. In fact, why are some solvents used preferentially over others? The answer to this question is quite simple: when chromatography was still young, these liquids had to have several important features. They could not be so viscous that they would not flow through packed columns by gravity or with the application of modest pressure. They could not be so volatile that they would evaporate before reaching the column outlet. In short, they had to be well-behaved liquids at ambient conditions. Chester further states that this thinking has carried through to the present day and he emphasizes that unified chromatography begins with the simple notion of rejecting ambient temperature and outlet pressure as being necessarily correct. I can really recommend the chapter in the aforementioned book to gain new insights into the concept of unified chromatography. Then it becomes clear that high-temperature HPLC or ultra high-pressure HPLC are only different variants of the same separation principle.

At this point I would like to direct the discussion to speed. In Chapter 1, the concept of the Height Equivalent to a Theoretical Plate (HETP) was introduced on a very simple basis. The preliminary conclusion was that as the temperature is increased, the operator does not have to worry about a loss in efficiency when the analysis is run in the C-term dominated region of the van Deemter curve. As the temperature is increased, the C-term "flattens out", which means that the curve does not increase much when the flow rate is increased at elevated temperature. This was highlighted by the data given in Figure 1.1. It was also discussed that the optimal velocity of the mobile phase to minimize the theoretical plate height shifts to higher linear flow rates. In effect, this will lead to the paradoxical situation that although the pressure drop across the column at a constant linear velocity will drop, the pressure to run the column at the optimal linear velocity will increase.

A very nice graphical as well as theoretical treatment can be found in a publication by Nguyen and co-workers, which is shown in Figure 4.12.[14] Here, the plate height H as a function of the linear velocity u is depicted for three different temperatures.

It can be clearly seen that as the temperature is increased, the optimal linear velocity shifts to higher values. If the separation is carried out at 30 °C, the minimum in the $H(u)$-curve is obtained when the linear velocity is adjusted to 5.1 mm s^{-1}. As the temperature is then increased to 60 °C and 90 °C, the optimum linear velocity increases to 7.9 mm s^{-1} and 12 mm s^{-1}, respectively. Therefore, if you would like to always operate the column at the minimum of the $H(u)$-curve, the flow rate has to be adjusted accordingly. This also means that the proposition that when the temperature is increased the pressure will be lower can be somewhat misleading. In fact, at very high temperatures it might

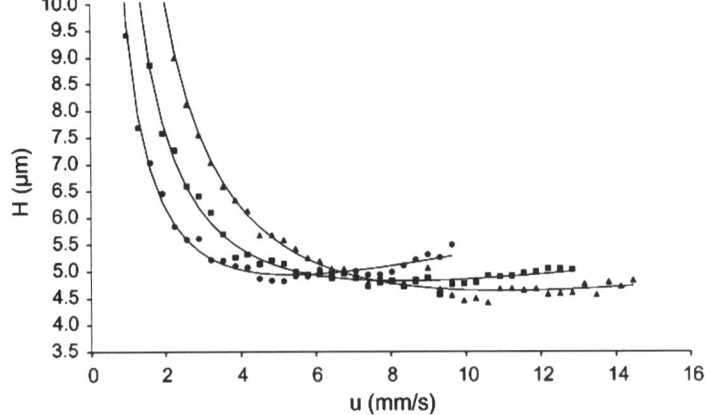

Figure 4.12 Plate height (H) as a function of the linear velocity (u) for each of the tested temperatures: ●, 30 °C; ■, 60 °C; and ▲, 90 °C.[14] Fitting of experimental data was performed with the Knox equation: $H = A \cdot u^{1/3} + B/u + C \cdot u$. (Reproduced with kind permission from Elsevier.)

Mobile Phase Considerations

well be that the pressure which is necessary to operate the column at the optimum conditions is much higher than at ambient temperature. This is often not considered, because at that point the chromatographic theory seems to be extremely counterintuitive to common knowledge. However, a column can be also operated around this van Deemter minimum without a significant loss in efficiency as is evident from Figure 4.12. But now I would like to derive this conclusion from the van Deemter equation, which should be known to every experienced chromatographer:

$$H_u = A + \frac{B}{u} + C \cdot u \tag{4.2}$$

More specific, the van Deemter equation can be written in a way to demonstrate the dependence of $H(u)$ on the particle diameter (d_p) and the diffusion coefficient of the solute in the mobile phase (D_M):

$$H_u = A' \cdot d_p + B' \cdot \frac{D_M}{u} + C' \cdot \frac{d_p^2}{D_M} \cdot u \tag{4.3}$$

From this expression, already some fundamental aspects can be derived. First of all, to obtain an absolute increase in column efficiency, which corresponds to a decrease in the plate height, the particle diameter should be reduced. This has been achieved by the introduction of sub 2 µm particles and the technique which is known as ultra high-pressure liquid chromatography is now widely used in routine laboratories. Although it might now be assumed that the absolute plate height can also be decreased with increasing temperature, numerous studies have not confirmed this suggestion.[14–17] I will come back to this question in Chapter 6, where I explain the impact temperature has on retention, selectivity and efficiency. I will then give more insight into this highly debated question and you will see that the influence of temperature on efficiency is a subject of much controversy in the scientific literature. Also, the experimental results, which have been gathered by many authors, can be interpreted very ambiguously. So for the moment I will leave this question of how temperature affects efficiency unanswered, but will focus instead on the effect of speed of separation when increasing temperature, where the reduction in viscosity is of utmost importance.

In order to evaluate the effect of temperature on the theoretical gain in separation speed, we have to take a closer look at the diffusion coefficient D_M of the solute in the mobile phase. According to Wilke and Chang, the diffusion coefficient can be written as:

$$D_M = 7.4 \cdot 10^{-8} \frac{\sqrt{\Psi_2 M_2}}{\eta V_1^{0.6}} T \tag{4.4}$$

where T is the absolute temperature, M_2 is the molecular weight of the solvent, V_1 is the molar volume of the solute and η is the viscosity of the mobile phase.

Ψ_2 is the association factor for the solvent, which is generally assumed to be 1 for non-polar solvents, 1.9 for methanol and 2.6 for water.[18] If it is assumed that the molar volume of the solute is not influenced by temperature, the diffusion coefficient is directly proportional to the temperature and inversely proportional to the viscosity of the mobile phase, which is also a function of temperature. If you have a closer look at Equation (4.5), it can be derived that as the temperature is increased, the viscosity will drop for a given mobile phase and thus, the diffusion coefficient of the solute in the mobile phase will increase. This can be written as:

$$\frac{D_M \cdot \eta}{T} = \text{const} \tag{4.5}$$

provided that the molar volume, the molecular weight and the association factor are not a function of temperature. What does this mean? Principally, if the diffusion coefficient of a solute at a given temperature is known, the diffusion coefficient at any other temperature can be easily calculated, provided the viscosity of the mobile phase is known. Also, the gain in speed, which is related to the ratio of temperature and viscosity, can be calculated. Thus, to operate a column at the optimum linear velocity, the flow rate needs to be increased. For all solvent systems listed in Table 1.1, this theoretical gain factor was calculated based on the comparison of viscosity at 25 °C with that at 250 °C. It is remarkable that quite large "speed" factors result, which are around 40 for the water–ethanol system and around 50 for the water–isopropanol system.[2] It is not surprising that the highest gain factors result for the higher chain alcohols because these mixtures run through an extremely large viscosity maximum at 25 °C.

In order to evaluate the practical benefit of temperature on separation speed, a dedicated HPLC experiment was performed in order to answer the question of whether a high gain factor in speed can be obtained on a system with a pressure limit of 1200 bar. Please note that the data was acquired on a system with an enlarged pressure range of up to 600 bar (Agilent 1200 Rapid Resolution). For this experiment, three columns were used and the maximum pressure drop for a linear solvent gradient was recorded, which can be found in Table 4.1. Here, the pressure maximum is shown for all solvent systems at three different temperatures. The temperature was adjusted to 40 °C, 70 °C and 120 °C, respectively. An upper temperature limit of 120 °C was chosen because most stationary phases will degrade rapidly if a higher temperature is selected, as will be shown in Chapter 5.

It is obvious that at a temperature of 120 °C, the highest pressure was observed for the water–isopropanol system, which exhibits a very pronounced viscosity maximum. Using the XBridge column with a particle diameter of 2.5 µm, the highest pressure during the gradient run was 258 bar, while for the Zorbax SB column the pressure was 306 bar. The results were in good agreement with the theoretical expectations, where the column packed with the smaller particles will yield a higher pressure, provided that all other parameters

Mobile Phase Considerations

Table 4.1 Dependence of system pressure on temperature and mobile phase for different column dimensions.

Mobile phase	$T/°C$	Waters Xbridge 5 cm, 2.5 μm, 3.0 mm : P/bar ($F/ml\,min^{-1}$)	Agilent Zorbax 5 cm, 1.8 μm, 3.0 mm : P/bar ($F/ml\,min^{-1}$)	ZirChrom-PBD 15 cm, 5 μm, 3.0 mm : P/bar ($F/ml\,min^{-1}$)
water–methanol	40	414 (2.9)	553 (2.2)	274 (4.4)
	70	266 (4.5)	353 (3.4)	185 (6.5)
	120	171 (7.0)	218 (5.5)	126 (9.5)
water–acetonitrile	40	273 (4.4)	365 (3.3)	187 (6.4)
	70	192 (6.3)	250 (4.8)	134 (9.0)
	120	131 (9.2)	165 (7.3)	95 (12.6)
water–ethanol	40	> 600 (-)	> 600 (-)	not possible
	70	367 (3.3)	442 (2.7)	250 (4.8)
	120	218 (5.5)	259 (4.6)	164 (7.3)
water–THF	40	485 (2.5)	> 600 (-)	not possible
	70	320 (3.8)	405 (3.0)	214 (5.6)
	120	227 (5.3)	270 (4.4)	167 (7.2)
water–isopropanol	40	> 600 (-)	> 600 (-)	not possible
	70	450 (2.7)	558 (2.2)	not possible
	120	258 (4.7)	306 (3.9)	204 (5.9)
water–acetone	40	367 (3.3)	467 (2.6)	245 (4.9)
	70	242 (5.0)	298 (4.0)	168 (7.1)
	120	160 (7.5)	188 (6.4)	117 (10.3)

are left unchanged. On the basis of the data presented in Table 4.1, at a constant flow rate of 1 ml min^{-1} it is possible to calculate the highest flow rate which could be achieved provided that a system with a pressure limit of 1200 bar is used. This calculation is based on the assumption that the pressure is directly proportional to the flow rate.[vi]

It is obvious that binary systems of water and an alcoholic modifier can be very troublesome at low temperatures. Note that the maximum pressure of the HPLC system, which was employed to measure the pressure at a flow rate of 1 ml min^{-1}, was already exceeded at a flow rate of 1 ml min^{-1} when a linear gradient was run at 40 °C for the mixtures water–ethanol and water–isopropanol. At this temperature, an HPLC system with a pressure limit around 1000 bar is needed to work at 40 °C when the flow rate is adjusted to 1 ml min^{-1}. However, increasing the temperature to 70 °C means that the pressure dropped below 600 bar for all the solvent systems. Based on these numbers it can be shown that at 120 °C, a tremendous gain in speed might be obtained on all three supports if an ultra high-pressure system is used. For water–acetonitrile, the flow rate might be increased to 9.2 ml min^{-1}, while for

---

[vi] Please remember that at very high pressures, deviations from this linearity due to frictional heating might occur as was already explained in Chapter 3. However, for the purpose of a rough estimation, these effects can be neglected.

the water–isopropanol system the flow rate can only be increased to 4.7 ml min^{-1}. However, as can be derived from the viscosity data which are compiled in Appendix B, at a temperature of 250 °C, the maximum flow rate which can be achieved on a given column should be identical regardless of the solvent system.

Although a really high gain in speed might be obtained, the flow rates to achieve this speed factor are very high. Currently, no commercially available HPLC system is capable of generating a flow rate above 5 ml min^{-1} at a maximum pressure of about 1200 bar. The Acquity system from Waters is only able to deliver a flow rate of about 1 ml min^{-1} at the highest operating pressure, while the Infinity system from Agilent is able to deliver flow rates up to 2 ml min^{-1} at a maximum pressure of 1200 bar. When the flow rate is increased further, the maximum pressure these systems can deliver is lower. Although the Infinity system is able to generate a flow rate of 5 ml min^{-1}, the maximum pressure is only 600 bar at the highest flow rate. From a purely theoretical standpoint, this is counterproductive, because it would be highly desirable that the maximum pressure can be maintained over the whole flow rate range. However, if you have a closer look on the van Deemter curves in Figure 4.12, it is obvious that the column can also be operated in the B-term region with only a minimal loss in efficiency. Note that the optimum flow rate at 90 °C is 12 mm s^{-1}, but it is undoubtedly possible to operate the column at 6 mm s^{-1}. Clearly, from a purely theoretical standpoint, it is desirable to avoid operating the column in the low flow rate B-term dominated region, because there is not only a loss in efficiency but also a loss in speed. However, when the maximum available pressure of the HPLC system is too low to enable the column to be used with the optimum flow rate, this is still acceptable. In some cases it is necessary to reduce the flow rate even if the pressure of the HPLC system is not the limiting factor. When a mass spectrometer is hyphenated to a high-temperature HPLC system, it is not advisable to increase the flow rate above 0.5 ml min^{-1}, because the sensitivity strongly decreases for electrospray ionization. So what does this mean for high-temperature HPLC? The best way to obtain a high speed factor and to be compatible with a mass spectrometric detector is to decrease the diameter of the column, which means that the optimum flow rate will be lower. In my opinion, this strategy is the best option for the hyphenation of liquid chromatography with mass spectrometry, although it should be considered that a system with a very low extra column volume is needed for columns with an inner diameter below 2 mm. Principally, the "new" ultra high-pressure systems definitely fulfil this requirement, but there are other reasons which currently prevent many practitioners from using capillary LC instead of conventional HPLC. I will resume this discussion in Chapter 9, where I try to give a critical outlook on future developments, including the prospects of high-temperature HPLC in combination with capillary and nano LC.

But now I will continue to explain the effect temperature has on the polarity of the mobile phase. This is the third parameter we have studied in detail and I will show that the influence of temperature on the polarity of different solvent systems can be quite pronounced if a large temperature interval is considered.

Mobile Phase Considerations

4.3 Influence of Temperature on Static Permittivity

The static permittivity (DEC) is a measure of the polarity of a substance and like vapour pressure and viscosity, it is strongly dependent on temperature.[vii] In general, static permittivity decreases when temperature increases. Especially with water, the decrease in static permittivity is responsible for the break up of hydrogen bonds. Figure 4.13 shows the dependence of the static permittivity of water on temperature, as it is increased from 0 °C to 374 °C.[4]

As the figure shows, the decrease in static permittivity closely resembles the decrease in the viscosity of a pure compound, which means that it is not linear but exponential. Please note that the static permittivity of pure water around ambient temperature is very high when compared with that of methanol or acetonitrile. This is the reason why water is such a weak solvent in reversed-phase HPLC. The strong hydrogen-bond network means that most organic solutes will strongly adhere to the non-polar surface of a bonded support of a reversed-phase packing material. By increasing the concentration of the organic co-solvent during a solvent gradient, the elution strength of the mobile phase is increased which means that the time the solute spends in the stationary phase can be greatly reduced. Now it is clear why a temperature gradient has the same effect as a solvent gradient in RP-HPLC. By increasing the temperature, the static permittivity of the mobile phase is reduced. Hence, the solvent strength of the mobile phase is increased even if the concentration of the organic co-solvent is kept constant.

In the scientific literature, most studies on high-temperature HPLC have focused on a pure water mobile phase.[19–23] However, the same rules apply for

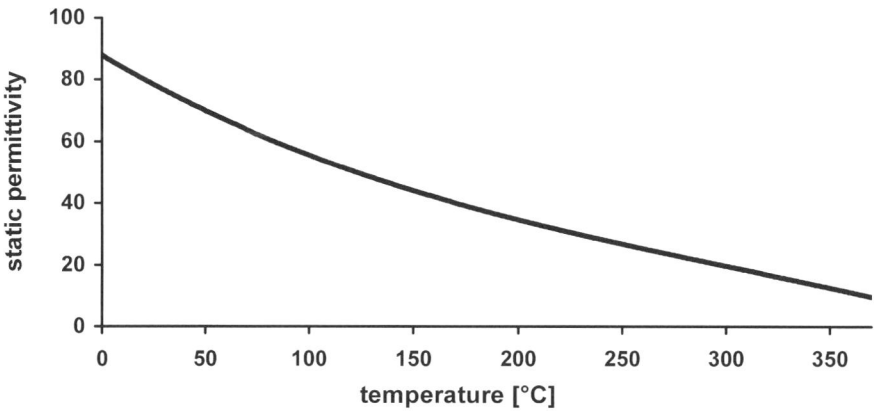

Figure 4.13 Dependence of static permittivity on temperature for liquid water between 0 and 374 °C. (Calculated with parameters from ref. 4.)

[vii] I would like to point out that in literature often the old term "dielectric constant" instead of "static permittivity" is found. But from a physical point of view this is not acceptable because the static permittivity of a solvent or a mixture is always a function of pressure and temperature.

binary mixtures composed of water and an organic co-solvent, which means that the elution strength of the mobile phase will always increase when either a solvent gradient is applied or the temperature is increased. In this respect, this is different when compared to the vapour pressure or the viscosity of binary mobile phases, as I have shown in the previous sections. For example, vapour pressure always runs through a maximum when a solvent gradient is applied and temperature is increased. Also, viscosity runs through a maximum when the composition of the mobile phase is changed, which can be very large at ambient temperature. However, for very high temperatures above 200 °C, this viscosity maximum can completely disappear. In contrast to this, the static permittivity of binary solvent mixtures always decreases when either the concentration or the temperature is changed. This phenomenon will be observed for all solvent systems listed in Table 1.1. I will not go into detail about the measurement technique, which can be found in a recent paper.[3] In this section, I will only focus on selected data to explain the effect temperature has on the static permittivity of pure solvents and binary solvent mixtures. The complete data sets can be found in Appendix C.

Figure 4.14 shows the static permittivities of the pure solvents, acetonitrile, acetone, isopropanol, ethanol, methanol and tetrahydrofuran, as a function of temperature at a constant pressure of 100 bar. As becomes obvious, the static

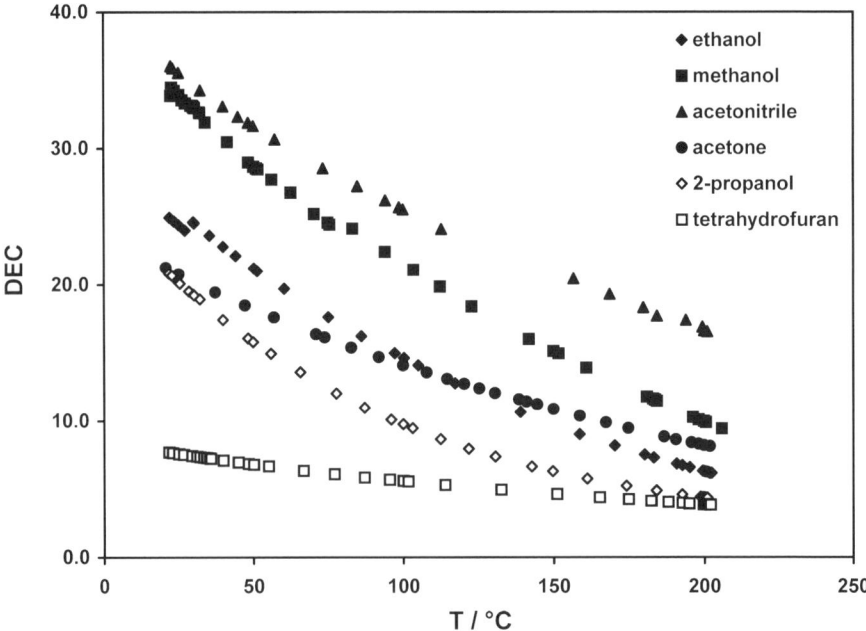

Figure 4.14 Temperature dependence of the static permittivities of pure solvents at a constant pressure of 100 bar.[3] (Reproduced with kind permission from Elsevier.)

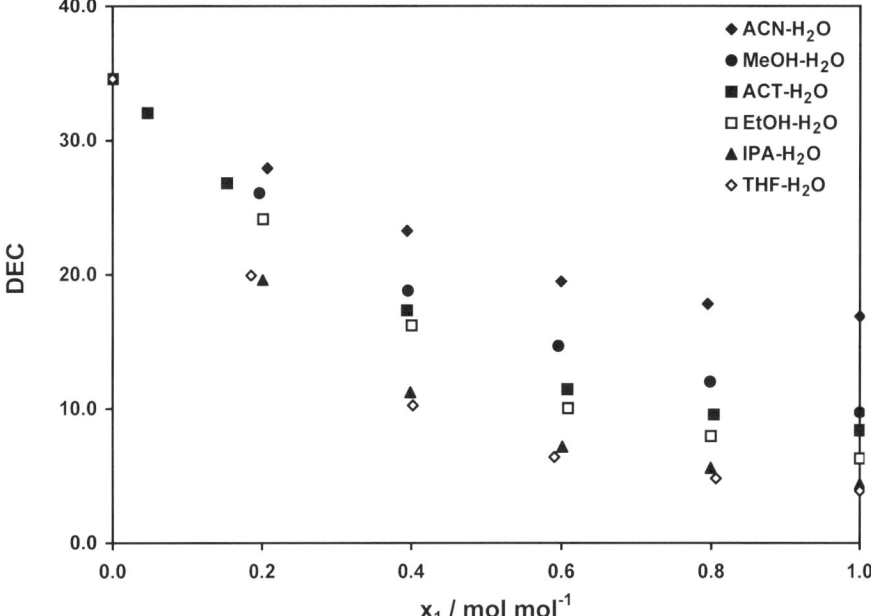

Figure 4.15 Experimentally determined static permittivities of binary solvent mixtures at 200 °C and 100 bar.[3] (Reproduced with kind permission from Elsevier.)

permittivity decreases for all solvents when the temperature is increased from 20 °C to 200 °C. This means that the polarity of these solvents decreases when the temperature is increased.

This is also true if you look at the data depicted in Figure 4.15. Here, the dependence of static permittivity on temperature is shown for binary mixtures at a temperature of 200 °C. The same effect is noticed as for the pure solvents, which means that static permittivity decreases as the temperature is increased. It can be concluded that this is a general phenomenon which is not only true for *e.g.* a pure water mobile phase but for all the binary mobile phases described here.

The question might now be asked as to what is the benefit of this phenomenon in liquid chromatography? At this point it is only important to keep in mind that this effect can be used for method development and plays a very important role when special hyphenation techniques are considered. If only the static permittivity of water at elevated temperature is considered, then many authors have concluded that water at temperatures around 250 °C has a comparable elution strength to methanol or acetonitrile at ambient temperature. However, the interactions of a solute between the mobile and stationary phase are far more complex. It is not admissible to reduce the elution strength of the mobile phase to a single parameter. In nearly all textbooks about HPLC, the eluotropic series can be found which classifies solvents according to their polarity or elution strength. Often, the sequence of solvents is based on the polarity parameter, which defines the polarity of a specific solvent relative to other solvents.[24] The elution strength is once again reduced to a single parameter, neglecting the various interactions which can take place between the analytes, as well as the mobile and stationary phases. As the data which are compiled in Appendix C reveal, the polarity of the different binary solvent systems strongly depends on temperature. While the eluotropic series strictly defines the sequence of the pure solvents at ambient temperature, the data we have published show that there are also changes in the relative order of the polarities as the temperature is increased. It is also interesting to note that the two solvents, methanol and acetone, have the same polarity parameter at room temperature which is equal to 5.1.[24] This should also mean that the respective mixtures of water with these solvents will always have the same elution strength. However, if the static permittivities of these two solvents are compared, it is evident that they exhibit different polarities. This can be also derived from chromatographic experiments where acetone is usually the stronger eluent when compared with methanol.

In order to visualize the solvent strength, an experiment was performed where a test mixture containing thirteen derivatized aldehydes and ketones was separated on a Zorbax SB column using a linear gradient. Please note that the flow rate has been kept constant in order to show the difference in elution strength for the chosen solvent systems. As can be clearly seen from the chromatograms in Figure 4.16, the analysis time can be decreased by using different solvent systems. Because all other parameters have been kept constant, the analysis time is only influenced by the nature of the organic co-solvent. As could be anticipated, by using the longer chain alcohols ethanol and isopropanol instead of methanol, the retention of all compounds is decreased while the resolution and selectivity of the separation remains virtually unchanged. This means that a remarkable reduction in the overall analysis time can be achieved by choosing a solvent system with similar molecular interactions but higher elution strength.

The resolution for a given peak pair can also be quite markedly influenced by changing the selectivity of the separation. As becomes evident from Figure 4.16, there are two groups of critical peak pairs, marked by a star and an asterisk. The first group contains three peaks while the second group contains four peaks. Using methanol, the first critical peak pair is only partially resolved, which is also true for the second group of critical peak pairs. In addition, the overall run time is about 20 minutes and therefore the longest run time. When ethanol or isopropanol are used instead of methanol, the resolution for these peak pairs is even worse. The same is true for acetonitrile and tetrahydrofuran. The best resolution is obtained for a binary mobile phase consisting of water and acetone. I would like to point out that the screening for the best solvent system can be easily accomplished without changing the stationary phase. If, for a given separation problem, the optimal solvent system has been found, the separation can be accelerated by adjusting the flow rate.

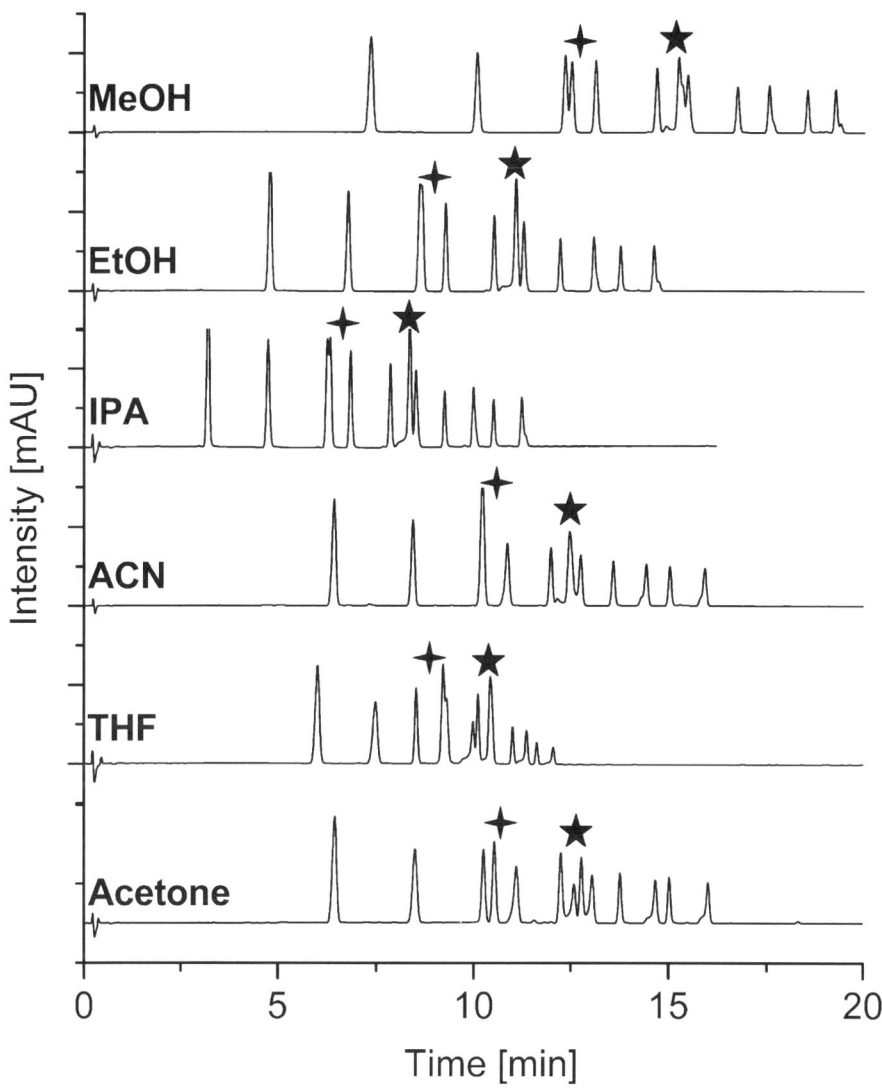

Figure 4.16 Chromatogram of a test mixture containing thirteen derivatized aldehydes and ketones. Chromatographic conditions: column: Agilent Zorbax SB C-18 (5 cm × 3.0 mm ID; 1.8 µm); solvent A: water; solvent B: organic co-solvent (see individual chromatograms); solvent gradient: 5 to 100% solvent B in 30 min; flow rate: 1.0 ml min^{-1}; temp.: 70 °C; detection: UV at 360 nm.[3] The star and the asterisk mark two groups of peaks which consist of three and four analytes, respectively. (Reproduced with kind permission from Elsevier.)

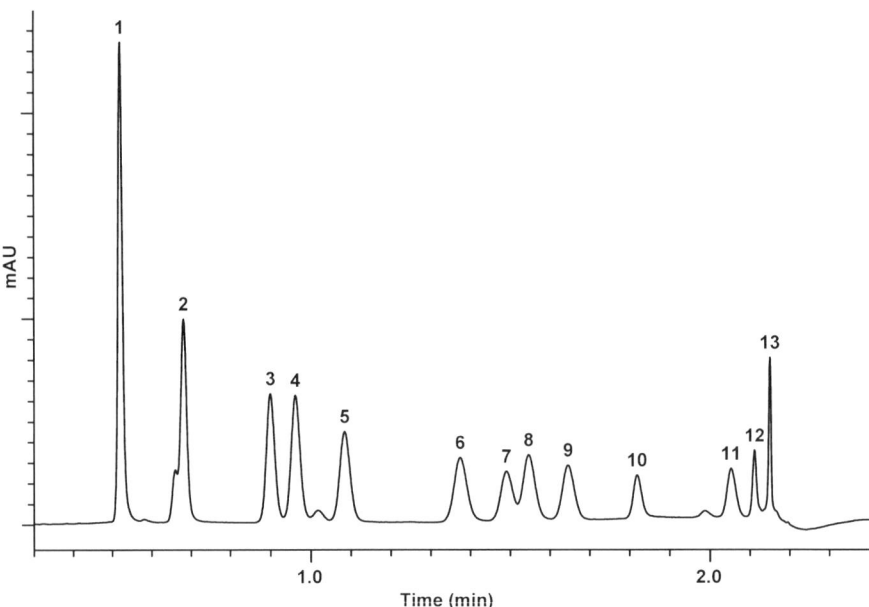

Figure 4.17 Optimized method for the separation of thirteen derivatized aldehydes and ketones with Waters Acquity UPLC at 64.6 °C. Chromatographic conditions: mobile phase: water–acetone, with solvent gradient elution; flow rate: 1.0 ml min^{-1}; injection vol.: 1.0 µl; max. pressure during solvent gradient: 786 bar; detection: UV at 360 nm.

Figure 4.17 shows a method where the analysis time and resolution have been optimized with the help of computer optimization software (DryLab 2000 Plus). As can be seen, the analysis time is reduced to 2.2 minutes at a temperature of 64.6 °C.

Please note that it was not possible to increase the flow rate further, because at flow rates above 1 ml min^{-1}, the maximum pressure that the system could generate was reduced. Using a high-pressure system with no pressure-dependent flow restriction would therefore allow the analysis time to be speeded up even further.

At this point, I would like to summarize some important facts which should be kept in mind. *By increasing the temperature, the static permittivity and hence the "polarity" of all solvent systems can be reduced. This is a general phenomenon and is not restricted to pure compounds. A complete substitution of organic solvents by pure (buffered) water is possible in reversed-phase HPLC*, enabling some very interesting hyphenation techniques, which will be described in Chapter 8. However, before I close this chapter, a very unusual behaviour needs to be explained. This relates to the water–tetrahydrofuran system, which is often used to adjust the selectivity of a separation along with water–methanol and water–acetonitrile.

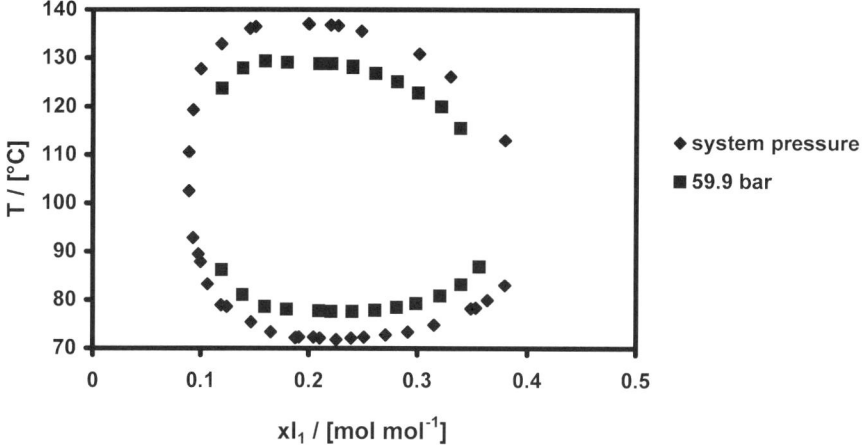

Figure 4.18 Liquid–liquid phase equilibrium of tetrahydrofuran (1)–water (2). (See text for further details.)

4.4 The Water–THF System

As I have shown so far, for most solvent systems a steady increase in temperature will only lead to continuous changes in the respective properties. This holds true for vapour pressure, viscosity and static permittivity. One key aspect is that in liquid chromatography, the components of the mobile phase which are mixed should be fully miscible in any concentration. Otherwise, problems can arise because of the immiscibility of the mobile phase components. At a first glance, all solvent systems which are studied in this book are fully miscible in any concentration. However, this is not true for the water–tetrahydrofuran system which deserves special attention. As the temperature is increased above 70 °C, the two solvents start to demix because the system is characterized by a miscibility gap, which is pressure and temperature dependent. The extent of this miscibility gap is illustrated in Figure 4.18 at two pressures. The higher the pressure, the less pronounced the miscibility gap is. Above a critical pressure of 247 bar the miscibility gap is nonexistent.[viii]

This phenomenon can have serious implications for liquid chromatographic separations when temperature, pressure and concentration are in the range of the miscibility gap. Usually, a pressure gradient exists in the axial direction of the column. The inlet pressure can be as high as 1200 bar when a system is used which is designed for ultra high-pressure liquid chromatography. However, as the mobile phase moves through the packed bed, there is a steady decrease in the column pressure. At the column outlet, the pressure approaches atmospheric pressure, which is far below the critical pressure for the formation of the

---

[viii] Please note that in this context, "critical pressure" refers to the pressure above which the miscibility gap is nonexistent. This has nothing to do with the definition of the critical pressure of a supercritical fluid.

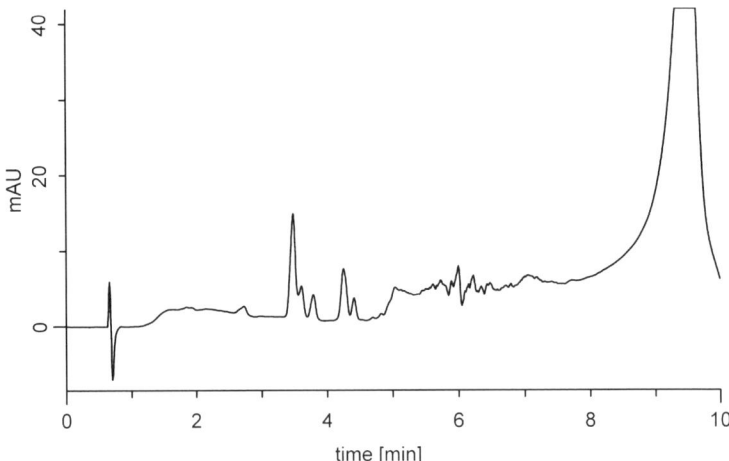

Figure 4.19 Separation of a test mixture containing thirteen derivatized aldehydes and ketones. Chromatographic conditions: column: Agilent Zorbax SB C-18 (5 cm × 3.0 mm ID; 1.8 μm); solvent A: water; solvent B: tetrahydrofuran; solvent gradient: 5 to 100% solvent B in 10 min; flow rate: 1.0 ml min^{-1}; temp.: 120 °C; detection: UV at 360 nm.

miscibility gap. Thus, even for separations which are carried out with a high inlet pressure, the miscibility gap within the column is likely to be present if the temperature is between 70 and 140 °C. In contrast to separations at ultra high pressures, conventional separations often do not require a very high pressure. Most liquid chromatographic separations can be carried out with an overall pressure below 250 bar. Therefore, the critical pressure is already below the threshold value for the formation of the miscibility gap from the point where the mobile phase is preheated. This means that the mobile phase might start to demix immediately when the temperature is between 70 and 140 °C. The question therefore arises as to what will happen with this mixture on the column when the temperature of the mobile phase is equal to or higher than the lower critical temperature of the miscibility gap? There is the assumption that depending on the experimental conditions, either a partial demixing or a preferential adsorption of THF will occur. Therefore, we were interested to investigate this phenomenon more deeply.

The chromatographic experiments were carried out in solvent-gradient mode using a binary mobile phase of water and tetrahydrofuran, where the concentration of THF was linearly increased from 0 to 100% over 30 minutes. The flow rate was adjusted to 1 ml min^{-1}. Again, a standard mixture containing 13 derivatised aldehydes and ketones was used. The first experiment was carried out at 70 °C and is included in Figure 4.16, while the second experiment was performed at a temperature of 120 °C which is within the miscibility gap. It can be clearly seen from Figure 4.16 that for a temperature of 70 °C, the peaks elute as symmetrical bands. The mixture is not completely resolved, but this is not important for this experiment. In contrast, at the higher temperature only a group of four partly resolved peaks is seen (see Figure 4.19). Other peaks cannot be detected, but the baseline rises and between 8 and 10 minutes a very large peak appears, probably containing the rest of the analytes. These results clearly demonstrate that the miscibility gap should be avoided. Otherwise, inconsistent chromatograms are obtained.

What is the consequence if you would like to work with the water–tetrahydrofuran system in high-temperature HPLC? Although this mixture exhibits a miscibility gap in the temperature range between 70 and 140 °C, it can be used as a mobile phase in high-temperature liquid chromatography. However, when a solvent gradient is applied and the chromatographic conditions are such that the miscibility gap may be approached within the column, inconsistent data might be produced. This means that retention times of test solutes cannot be reproducibly measured. Nevertheless, we did not observe an increase in the system pressure due to the build-up of two immiscible phases within the column as would have been expected. However, for the practical application of high-temperature liquid chromatography, the miscibility gap should be avoided. This means that the temperature has to be adjusted below 70 or above 140 °C when a solvent gradient is applied over the full concentration range. If it is necessary to adjust the temperature between 70 °C and 140 °C, the concentration of THF should always be higher than 40 mol%. In this case, the miscibility gap does not form regardless of the pressure and temperature in the column.

4.5 The Dortmund Data Bank

In this chapter, some very interesting data has been presented which is very useful for the practitioner. However, numerous people complain that it is really difficult to search for physicochemical data for these solvent systems over a large temperature interval. Indeed, much data has been published but as was already pointed out, a literature search can be quite time consuming. In addition, the data is often scattered which means that a full data set over a large temperature interval is not covered in one publication. In addition, these data were often published in physical chemistry journals which are mostly overlooked by the chromatographic community. Also, it is very difficult for a separation scientist to evaluate the quality of the data. Different techniques are often employed to measure some physicochemical properties and it is not easy to compare two different data sets.

A better way is to consult a database where all data has been collected. One of the world's biggest databases for pure compounds as well as mixtures is the Dortmund Data Bank (DDB). This databank was started in 1973 in the research group of Prof. J. Gmehling at the University of Dortmund with the compilation of VLE data for normal boiling points of mostly organic compounds.[25] It was later extended to include pure component properties, liquid–liquid equilibrium data, excess enthalpies, activity coefficients at infinite dilution, as well as data on various other thermodynamic and thermophysical properties. The Dortmund Data Bank is the largest data bank for experimental data on pure components and mixtures and available *via* in-house versions, online (STN, Internet) and in printed form. The Dortmund Data Bank provides routines which allow for a quick search of relevant information concerning pure components and mixtures. As of 2008, the DDB contains experimental data from 61 000 references from approximately 1800 journals for 28 600 compounds, including salts, adsorbents and polymers. Besides data from the open literature, a significant portion of experimental data originates from private communications, MSc and PhD theses, as well as from the chemical industry. DDB covers worldwide sources in all languages. Special conditions are available for the use of the DDB in research by individual research groups, whole departments or research institutions in a country.

In my opinion, it's a good starting point to first consult the DDB and search for available data sets before resources are spent on extensive experimental studies. When we started the research project on the determination of the vapour pressure, viscosity and static permittivity dependence on temperature for selected binary solvent mixtures, the DDB was searched for relevant information. It became obvious that it would be necessary to conduct some experiments in order to acquire the missing data sets. Especially for high-temperature HPLC, there was virtually no data on the temperature dependence of viscosity for binary eluent mixtures. For pure components, neither experimental data nor suitable thermodynamic models were available to make a precise prediction of the solvent properties. A completely new measurement technology had to be developed in order to acquire accurate data on the temperature dependence of the static permittivity of these solvent mixtures. The data is now accessible *via* a research report and three publications in the *Journal of Chromatography, A*.[1–3,9] Also, the data has been incorporated into the DDB.

Complete data sets are now available for the temperature dependence of the vapour pressure, the viscosity and the static permittivity of all binary eluent mixtures specified in Table 1.1. A large temperature interval from ambient temperature up to 200 or even 250 °C is covered. The quality of the data was also checked, either by applying appropriate thermodynamic models, like the Non Random Two Liquids (NRTL) or Predictive Soave-Robinson-Kwong (PSRK) model for the prediction of vapour–liquid equilibria, or by comparing the newly acquired data with literature data already stored in the DDB. This is very convenient because, with the availability of sophisticated software tools and prediction models, outliers can be easily identified and the error of the experimental measurements can be assessed.

The precise determination of this data can be of great value for all practitioners as well as instrument manufacturers. The practitioner will benefit from this data, because now he or she is able to design the HPLC system according to his or her special needs. This means that the user now has good data from which to decide if a back-pressure regulator has to be placed before or behind a detector, or if the hot mobile phase leaving the column has to be cooled down before it is introduced into the detector. Up until now, there has been much uncertainty on how to establish a system for high-temperature HPLC. In my opinion, this was mainly due to a lack of temperature-dependent data for important parameters. Also, the data can provide a good basis for the on-going discussion about the kinetic aspects in liquid chromatography.[15,26–35] In all these equations, viscosity is a key parameter for calculating the efficiency or the system pressure which results when a certain mobile phase at a defined temperature is used.

References

1. T. Teutenberg, P. Wagner and J. Gmehling, *J. Chromatogr., A*, 2009, **1216**, 6471.
2. T. Teutenberg, S. Wiese, P. Wagner and J. Gmehling, *J. Chromatogr., A*, 2009, **1216**, 8470.
3. T. Teutenberg, S. Wiese, P. Wagner and J. Gmehling, *J. Chromatogr., A*, 2009, **1216**, 8480.
4. Physikalisch Technische Bundesanstalt, *PTB-Stoffdatenblätter SDB 11*, Fachinformationszentrum Chemie GmbH, Berlin, 1995.
5. S. Cassel, P. Chaimbault, C. Debaig, T. Benvegnu, S. Claude, D. Plusquellec, P. Rollin and M. Lafosse, *J. Chromatogr., A*, 2001, **919**, 95.
6. T. Andersen, A. Holm, I. L. Skuland, R. Trones and T. Greibrokk, *J. Sep. Sci.*, 2003, **26**, 1133.
7. D. Guillarme, S. Rudaz, C. Schelling, M. Dreux and J. L. Veuthey, *J. Chromatogr., A*, 2008, **1192**, 103.
8. B. A. Ingelse, H.-G. Janssen and C. A. Cramers, *J. High Resolut. Chromatogr.*, 1998, **21**, 613.
9. T. Teutenberg, J. Gmehling and P. Wagner, Abschlussbericht Nr. 14514 N: *Entwicklung moderner Analysemethoden in der Flüssigkeitschromatografie durch Modulation der Temperatur und des Drucks von binären Lösungsmittelgemischen*, Arbeitsgemeinschaft industrieller Forschungsvereinigungen "Otto von Guericke" e.V., Köln, 2008.
10. H. Engelhardt and M. Jungheim, *Chromatographia*, 1990, **29**, 59.
11. U. D. Neue, E. Serowik, P. Iraneta, B. A. Alden and T. H. Walter, *J. Chromatogr., A*, 1999, **849**, 87.
12. L. C. Sander and A. W. Stephens, *J. Sep. Sci.*, 2003, **26**, 283.
13. T. L. Chester, in: *Multidimensional Chromatography*, ed. L. Mondello, A. C. Lewis and K. D. Bartle, John Wiley & Sons, Ltd, Chichester, 2002, p. 151–169.
14. D. T. T. Nguyen, D. Guillarme, S. Heinisch, M. P. Barrioulet, J. L. Rocca, S. Rudaz and J. L. Veuthey, *J. Chromatogr., A*, 2007, **1167**, 76.
15. D. Cabooter, S. Heinisch, J. L. Rocca, D. Clicq and G. Desmet, *J. Chromatogr., A*, 2007, **1143**, 121.
16. G. Vanhoenacker and P. Sandra, *J. Sep. Sci.*, 2006, **29**, 1822.
17. F. D. Antia and C. Horvath, *J. Chromatogr.*, 1988, **435**, 1.
18. C. R. Wilke and P. Chang, *AIChE J.*, 1955, **1**, 264.

19. R. M. Smith, *J. Chromatogr., A*, 2008, **1184**, 441.
20. R. M. Smith, *Anal. Bioanal. Chem.*, 2006, **385**, 419.
21. J. W. Coym and J. G. Dorsey, *Anal. Lett.*, 2004, **37**, 1013.
22. T. Teutenberg, O. Lerch, H. J. Götze and P. Zinn, *Anal. Chem.*, 2001, **73**, 3896.
23. R. M. Smith, R. J. Burgess, O. Chienthavorn and J. R. Bone, *LCGC Int.*, 1999, **17**, 938.
24. L. R. Snyder, *J. Chromatogr. Sci.*, 1978, **16**, 223.
25. http://www.ddbst.de/new/DDB.htm (last accessed October 2009).
26. F. Gritti and G. Guiochon, *J. Chromatogr., A*, 2009, **1216**, 4752.
27. A. de Villiers, F. Lynen and P. Sandra, *J. Chromatogr., A*, 2009, **1216**, 3431.
28. D. Cabooter, F. Lestremau, A. de Villiers, K. Broeckhoven, F. Lynen, P. Sandra and G. Desmet, *J. Chromatogr., A*, 2009, **1216**, 3895.
29. S. Heinisch, G. Desmet, D. Clicq and J. L. Rocca, *J. Chromatogr., A*, 2008, **1203**, 124.
30. D. Cabooter, F. Lestremau, F. Lynen, P. Sandra and G. Desmet, *J. Chromatogr., A*, 2008, **1212**, 23.
31. D. Cabooter, J. Billen, H. Terryn, F. Lynen, P. Sandra and G. Desmet, *J. Chromatogr., A*, 2008, **1204**, 1.
32. F. Lestremau, A. de Villiers, F. Lynen, A. Cooper, R. Szucs and P. Sandra, *J. Chromatogr., A*, 2007, **1138**, 120.
33. D. Clicq, S. Heinisch, J. L. Rocca, D. Cabooter, P. Gzil and G. Desmet, *J. Chromatogr., A*, 2007, **1146**, 193.
34. J. Billen, D. Guillarme, S. Rudaz, J. L. Veuthey, H. Ritchie, B. Grady and G. Desmet, *J. Chromatogr., A*, 2007, **1161**, 224.
35. G. Desmet, D. Clicq, D. T. Nguyen, D. Guillarme, S. Rudaz, J. L. Veuthey, N. Vervoort, G. Torok, D. Cabooter and P. Gzil, *Anal. Chem.*, 2006, **78**, 2150.

CHAPTER 5
Suitable Stationary Phases

Undoubtedly, the column is still the most important part of the whole chromatographic system. This is because a column has to be replaced more often than other hardware components. Therefore, the best column for a separation is not necessarily a column which gives the fastest separation. The batch-to-batch reproducibility of a column is a very strong criterion. If a method has been validated and is run in a regulated environment, it can be devastating to have to re-validate a complete method because the manufacturer has made some changes to the packing material. This has led to a widespread use of silica-based stationary phases whose bonding chemistry is well understood and which exhibit an excellent batch-to-batch reproducibility. Although silica-based phases are characterized by their high mechanical stability and excellent mass transfer properties, their robustness against aggressive pH and temperature conditions cannot compete with polymeric or metal oxide stationary phases. As has already stated by Neue, "A packing with all the advantages of silica, but with an expanded pH range is still the holy grail of HPLC".[1] When mobile phases are used with pH <2, the bonded phase is susceptible to hydrolysis, and at pH >8, particle erosion can occur due to the dissolution of the base silica particle. Moreover, elevated temperatures will accelerate the degradation.

Competitive materials for high-temperature HPLC include coated metal oxides like coated zirconium dioxide or titanium dioxide stationary phases.[2-4] Zirconia phases were introduced by Carr and co-workers some years ago. However, there is still a widespread reluctance to use these columns in industry since the retention mechanism is different from that of silica-based phases when ionic interactions are concerned.[5] In an excellent review, Nawrocki et al. pointed out that the chemistry of a zirconia surface is very complex and ion-exchange reactions play an important role.[4] Strong, hard Lewis acid sites, present on a zirconia surface, can interact with hard Lewis bases and these interactions are often considered troublesome. Therefore, silica-based stationary phases with an enhanced temperature range are needed to broaden the acceptance of high-temperature liquid chromatography as a routine method in industry.

Although high-temperature liquid chromatography attracts much interest, data on the stability of common stationary phases at high temperatures is limited. Some authors have published valuable information about the ruggedness of different types of columns at elevated temperatures and extended pH.[6-13] However, as yet a standardized protocol for a column ageing procedure at high temperatures does not exist. As a consequence the data generated by different authors is difficult to compare. In recent years our laboratory has developed several test procedures to monitor the stability of HPLC columns at high eluent temperature and extreme pH.

First of all I would like to explain, however, why it is necessary to develop and apply such a procedure. It is clear that column manufacturers often do not have data to support the specifications given in the respective data sheets for their columns. In fact, some manufacturers are over optimistic. When I was looking for new columns which might be used at high temperatures, I visited ANALYTICA in Munich in 2004, one of the biggest trade fairs for presenting new analytical technologies. At that time it was really hard to find a suitable column, so I was quite astonished when I read on a poster that a certain material could be used up to 250 °C. The vendor was sure that the column would be stable because he was convinced of his new technology. I could hardly wait to test the new column, but I was soon disappointed. After two or three columns had been tested, which all failed after the temperature was increased above 100 °C, the vendor hastily corrected the specifications. Clearly, this marketing strategy is successful as long as there are not enough people who will use a column at such high temperatures. Nevertheless, it is very dangerous because a new technique which relies on robust stationary phases can suffer if the user discovers that the specifications cannot be met in reality. What I have often observed is that when people are very enthusiastic about a new method or technique, they don't tell the whole story. Unfortunately, the scientific business is often not as objective as we would like it to be and marketing intrudes. Lectures given by scientists are often not much different from the presentations of marketing specialists from big companies: everything is excellent. If you don't obey by this rule, you will get no money from government agencies for funding and there is no cooperation from instrument manufacturers who are not interested in fundamental research. Nevertheless, I think that the goal of scientific research should be to report objectively about the facts which are gathered from experiments. Therefore, telling the truth about the pros and cons of different stationary phases is more helpful for the practitioner, because he or she will inevitably find out if a report is useful. However, the results reveal that a lot needs to be done to improve column stability as will become clear when you read this chapter.

Now, let's summarize what is the difficulty in designing columns which can be used for high-temperature HPLC. Ideally, there should be no influence of temperature on the stability of the packed bed over a long period. However, as we all know, by increasing the temperature, degradation is facilitated. Up until

Suitable Stationary Phases   89

now, there is absolutely no material which does not show some sign of degradation at high temperatures. Even polymer-coated metal oxide stationary phases display these effects. It is important to distinguish between two degradation pathways. First of all, there is the loss of the bonded phase or the support material by abrasion or dissolution. Especially silica-based materials are very prone to hydrolysis. Therefore, a better cross-linking of the bonded phase to the silica is desirable. The other problem comes up if the stationary phase is heated up and cooled down repeatedly, especially when temperature programming is applied. In this case, the hardware as well as the stationary phase are both subjected to expansion and contraction. This might result in the rearrangement or crushing of the packing within the column or the creation of voids, which are responsible for extra column band broadening.[i,14]

Therefore, the test procedure should reveal which columns are suitable for high-temperature HPLC.

Before I continue I would like to stress that this chapter is not intended to cover the complete retention mechanism for every type of support. Instead, I will focus on the most frequently asked question, "which columns can be used for high-temperature HPLC?" In every section, I will make references which should be consulted for further reading. Although it is a little bit out of date, the book from Neue on HPLC columns is an excellent introduction to column technology and should be consulted for further reading.[1]

5.1 Column Bleed

The degradation of a column is not usually an abrupt process. Rather, it is a continuous change in the column's properties. If you have already made some experiments at high eluent temperatures, you should carefully re-examine your data. If you have used a UV detector, you might observe that the baseline rises and the noise increases once a certain temperature is exceeded. The reason is that either hydrolysis of the support material or loss of the bonded phase is taking place. This phenomenon has been termed "column bleed" and results from a continuous wash-out of particulate matter. A very simple test procedure has been applied in our laboratory to compare different columns in terms of column bleed. Here, the columns are subjected to a temperature gradient. The temperature programme consists of an isothermal hold-up step for 5 minutes at 30 °C. After this, the temperature is raised over 5 minutes from 30 to 200 °C and is then held constant for 10 minutes. Afterwards, the system is cooled down to

---

[i] In a very exhaustive review about the limits of the separation power of one-dimensional column liquid chromatography, Guiochon has calculated the thermally induced expansion of a stainless-steel column, which supports the hypothesis of a particle rearranging due to high temperatures. The thermal expansion coefficients of most grades of stainless steel are of the order $1 \times 10^{-5}\,°C^{-1}$. The effect of temperature on the column length and its internal diameter may be negligible for a change of 10 °C. However, when the temperature is increased to *e.g.* 200 °C, the length of a 30 cm long column increases by 0.6 mm and its diameter by *ca.* 9 μm. This last value may seem to be negligible in a sense but it may result in quite significant effects when the column is packed with fine particles. If the column diameter expands by several particle diameters, the stability of the packed bed may be endangered.

30 °C and the baseline signal monitored for another 20 minutes. We used charged-aerosol detection, because this detector – like the evaporative light-scattering detector – should be ideally suited for the detection of particulate matter. The results of this experiment are given in Figures 5.1 and 5.2.

It is evident that the bleed of a "conventional" silica-based C-18 stationary phase is significantly higher than for columns which are designed for high-temperature HPLC. Nevertheless, these alternative materials are also prone to column bleed and hence, degradation of these columns will be also observed. I think that this experiment is quite simple and can be easily reproduced in your laboratory if you have a heating system which allows for temperature programming. However, it might be observed that for UV detection, a strong signal may result, which is due to the wash-out of strongly retained compounds which will be gradually eluted from the column. Therefore, the results need to be carefully interpreted and should be analyzed using two different detectors. The interested reader will find more information in a recently published paper, where the complete experiment is described in detail.[8]

I am sure that you will also encounter these effects which are typical for a column that is operated at elevated temperatures. If the baseline rises too much, then I would strongly recommend that you choose a lower temperature.

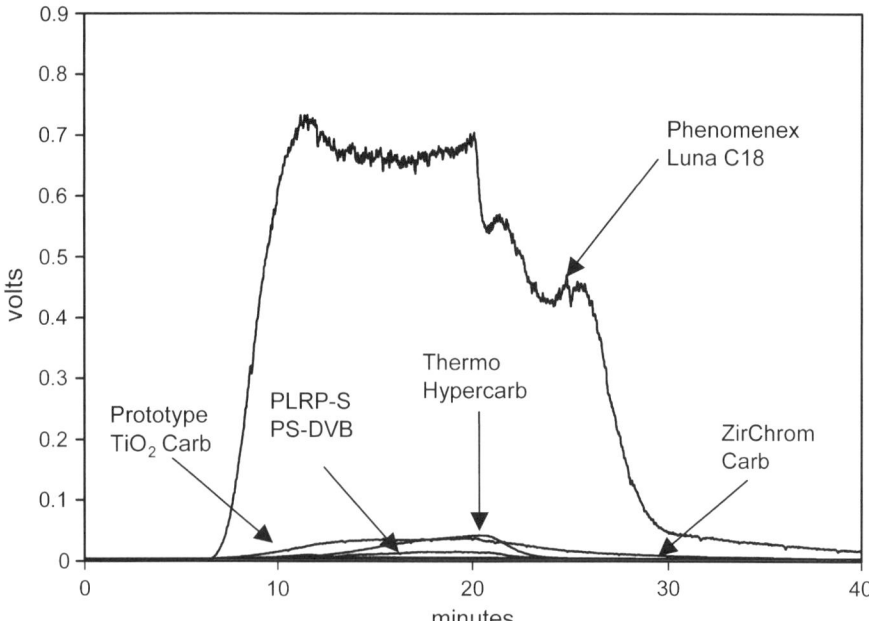

Figure 5.1 Dependence of detector response on temperature-programmed measurements for five different HPLC columns. Chromatographic conditions: detection: CAD; mobile phase: pure water; flow rate: 0.5 ml min^{-1}; temp. gradient: 30 °C for 5 min, then 30 to 200 °C in 5 min, then hold at 200 °C for 10 min, then cool to 30 °C.[8] (Reproduced with kind permission from Elsevier.)

Suitable Stationary Phases 91

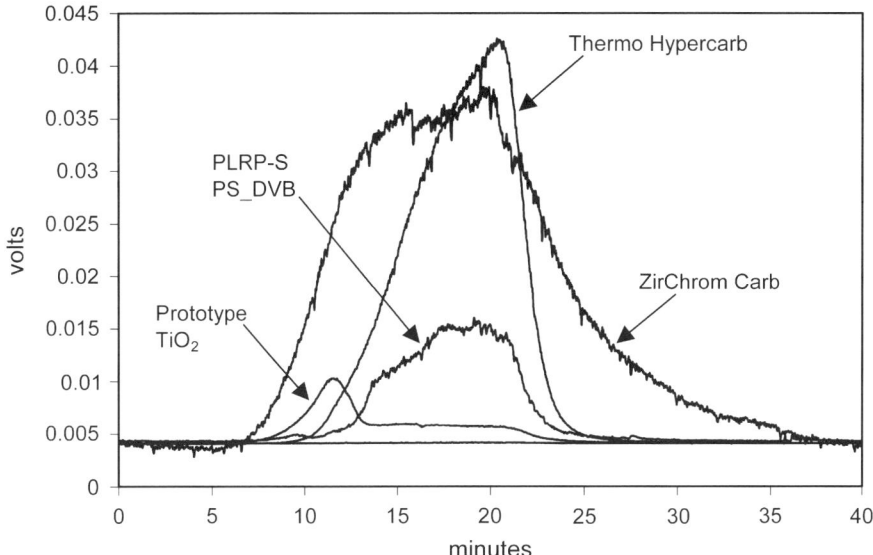

Figure 5.2 Dependence of detector response on temperature-programmed measurements for four different HPLC columns (C-18 column not included). Chromatographic conditions: detection: CAD; mobile phase: pure water; flow rate: 0.5 ml min^{-1}; temp. gradient: 30 °C for 5 min, then 30 to 200 °C in 5 min, then hold at 200 °C for 10 min, then cool to 30 °C.[8] (Reproduced with kind permission from Elsevier.)

Otherwise, the column might fail too quickly. Although this experiment will give you valuable information about the extent of column bleed, the procedure I will describe in the following section should monitor the loss in retention when a column is exposed to a hot mobile phase for a longer time interval.

5.2 Investigation of Column Degradation at High Temperatures

The following results that I would like to present are based on a test procedure we developed in our laboratory and which can be read in detail elsewhere.[9,10] I would like to describe the procedure here because we tested a lot of columns and also plan to test more columns in the future. I consider it important to mention the details of this testing protocol here, because other groups have also reported column tests in the scientific literature. Nevertheless, in some reviews or papers the reader will find information which is not based on hard facts. In many cases, only the specifications of the manufacturer are given without any experimental proof. Therefore, it is very difficult for a practitioner with no experience in high-temperature liquid chromatography to distinguish between real results and information which is only based on the manufacturer's data sheet. And as we have seen, this information should not be trusted blindly.

Our procedure is based on measuring the efficiency of a column with a standard test under defined conditions. Here, the Neue test was chosen because this test is well known and has been developed in order to characterize columns from different vendors.[15] When we test a column for its stability at high eluent temperatures it is not our primary interest to compare its performance to other columns or to measure the selectivity of a peak pair. The goal is to monitor the degradation of the column at defined intervals. We have termed these defined intervals "heating cycles". What does this mean? First of all, the column efficiency is measured for a brand new column as received by the manufacturer. Afterwards, the column is heated up to a defined temperature. For silica-based phases, we chose a temperature of 150 °C because in our opinion this is a very high temperature and columns which are not stable will degrade rapidly. The problem is that the test procedure is very time-consuming and therefore we applied very harsh conditions for the degradation test. The next issue is to find a suitable purging eluent. As I have outlined in Chapter 4.3, it is a general phenomenon that if the temperature is increased, the static permittivity of pure solvents as well as binary solvent mixtures decreases, which corresponds to a decrease in the eluent's polarity. Hence, organic co-solvents, which are normally needed to increase the elution strength of the mobile phase during a solvent gradient, can be replaced by water. Therefore, at high temperatures less organic solvent is needed to achieve the same retention for a certain compound when compared to ambient temperature. As is well known, water is a very aggressive solvent for silica-based columns. The higher the water content, the more rapidly dissolution of the silica gel or hydrolysis of the bonded phase from the support may occur. Therefore, we decided to use a very high content of water for the purging at high temperatures. For the investigation of silica-based materials, however, it might be necessary to add a small amount of an organic co-solvent. Otherwise, the column may fail after a few hours because of a phase dewetting, which has also been termed "phase collapse" by other authors.[16] Therefore, we used a binary mixture of water–methanol (90 : 10 v/v) for purging the column at high temperature. But before the column is purged at high temperatures, the initial performance is measured at the conditions of the Neue test. After this procedure, the mobile phase is changed to water–methanol (90 : 10 v/v) and is heated to 150 °C. Then, the column is flushed at 1 ml min^{-1} for exactly five hours. Afterwards, the column is cooled down to 25 °C and a solvent gradient is applied to remove hydrolyzed ligands. Then, the mobile phase is adjusted to phosphate buffer of pH 7–methanol (35 : 65 v/v), which corresponds to the conditions of the Neue test, and the performance of the column is measured again. This procedure is repeated five times, which we call a heating phase at neutral elution conditions. If the column is stable, the pH for the high-temperature purging is adjusted to 2.2. Now we can evaluate the effect of high eluent temperatures in combination with a low pH. Then, five heating cycles are run at acidic conditions and after each heating cycle, the performance of the column is measured. If the column even "survives" this treatment, then the pH is adjusted to 12.0 again using phosphate buffer and applying five heating cycles.

I hope that the general procedure is clear, because the chromatograms which are shown in the following sections are mainly based on that approach. I would also like to stress that there are many papers which deal with the investigation of column stability. However, in most cases it is not possible to directly compare the results of different authors, because different approaches were used, which should be kept in mind if someone states that a certain column is stable at 200 °C.

5.3 Silica-Based Stationary Phases

In this paragraph, I will give some examples of silica-based columns which appear to be very promising candidates for high-temperature HPLC. Although in the scientific literature many reports have been published where it is claimed that silica-based stationary phases are stable above 100 °C, it has not been possible to reproduce these observations in the laboratory of the author over the last ten years (see my comment in Chapter 2.2). I think that these were exactly the experiences which frustrated many groups who worked on high-temperature HPLC, and were responsible for the rapid loss of interest in this technique. If the column is not stable over a "reasonable" period, then this technique has no practical benefit. Figure 5.3 depicts a test chromatogram which was obtained after a brand new column was subjected to the test procedure described in the previous paragraph for exactly five heating cycles. The comparison of the test chromatogram before and after the fifth heating phase (as shown in Figure 5.3) is not very promising, revealing the hydrolysis of the silane bond.

Nevertheless, with the advent of hybrid silica technology, a significant improvement in terms of column stability has been achieved. The stabilization of the silica matrix *via* ethylene bridges can be regarded as the decisive breakthrough in column chemistry. Not only the stability against basic pH, which was the primary driving force of column manufacturers to improve their materials, but also increased stability towards high eluent temperatures has now been obtained. Figure 5.4 compares the stability of a modified packing material from the same manufacturer which was stabilized with ethylene bridges. It is obvious that there is only a minor shift in the retention of test analytes. In fact, the slight degradation is hardly noticed when the efficiency is compared before and after the high-temperature flushing at neutral elution conditions (see Figures 5.4a and 5.4b).

An even better result is obtained if a competitor column is used. Here, the efficiency of the column is completely maintained after the first five heating cycles, which can be seen from Figure 5.5b.

However, the interplay between temperature and pH can be devastating for the packing. Whilst column performance was only slightly affected after operation under acidic conditions and an additional heating time of 25 hours, an adjustment of the eluent to a basic pH completely "killed" the column. Nevertheless, the peaks were not totally lumped together and there was at least

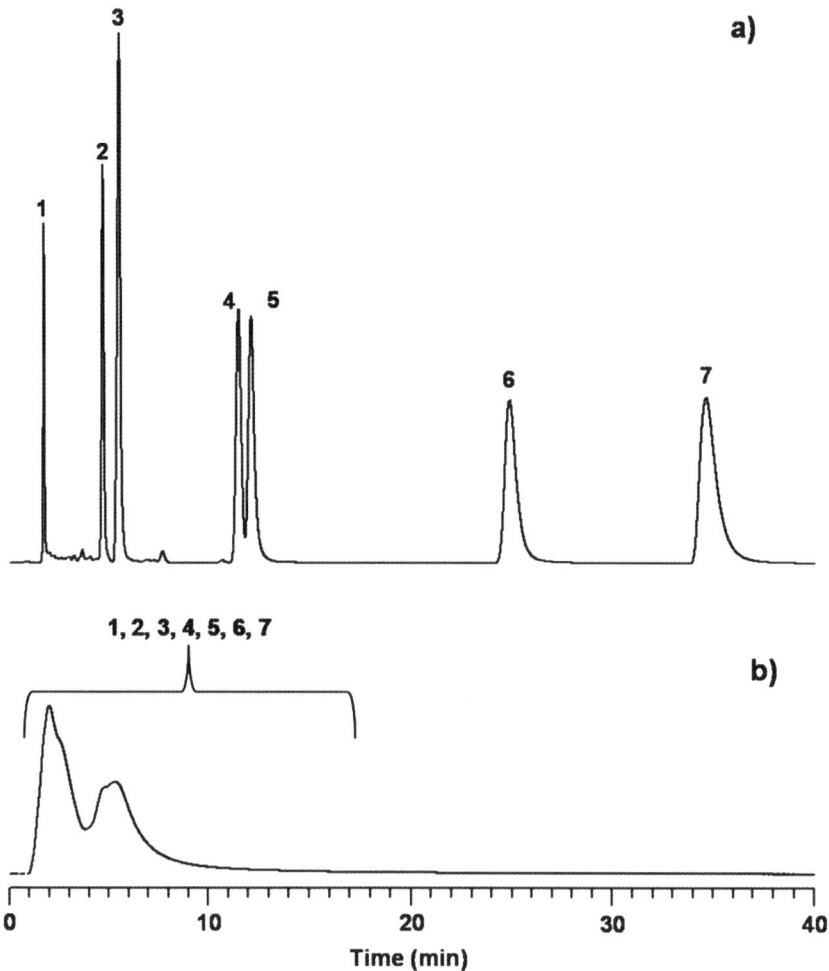

Figure 5.3 Temperature stability of a Phenomenex Gemini C18 column. Test chromatograms obtained (a) before the column was heated up to 150 °C, and (b) after the fifth heating cycle. Chromatographic conditions: mobile phase: phosphate buffer of pH 7–methanol (35 : 65 v/v); flow rate: 1 ml min^{-1}; detection: UV at 254 nm; temp.: 25 °C. Peaks: 1, dihydroxyacetone; 2, propyl paraben; 3, propranolol; 4, dipropyl phthalate; 5, naphthalene; 6, acenaphthene; and 7, amitriptyline.[10] (Reproduced with kind permission from Wiley-VCH Verlag GmbH & Co. KGaA.)

a partial separation of the test mixture. These examples serve to highlight that the modern phases are really much more stable than was the case ten or twenty years ago. Our results are also supported by other groups who have used this type of column when working in high-temperature HPLC.[17–24] These findings clearly demonstrate that with the availability of thermally rugged silica-based

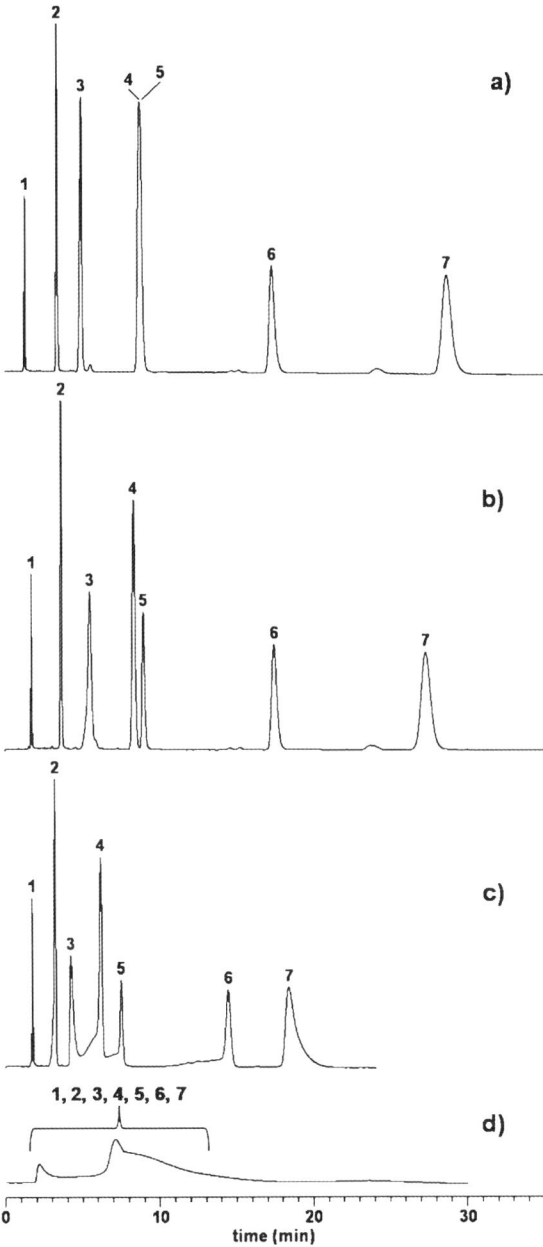

Figure 5.4 Temperature stability of a Phenomenex Gemini NX column. Test chromatograms obtained (a) before the column was heated up to 150 °C; (b) after the fifth heating cycle; (c) after the tenth heating cycle; and (d) after the thirteenth heating cycle. Chromatographic conditions: mobile phase: phosphate buffer of pH 7–methanol (35 : 65 v/v); flow rate: 1 ml min^{-1}; detection: UV at 254 nm; temp.: 25 °C. Peaks: 1, dihydroxyacetone; 2, propyl paraben; 3, propranolol; 4, dipropyl phthalate; 5, naphthalene; 6, acenaphthene; and 7, amitriptyline.[10] (Reproduced with kind permission from Wiley-VCH Verlag GmbH & Co. KGaA.)

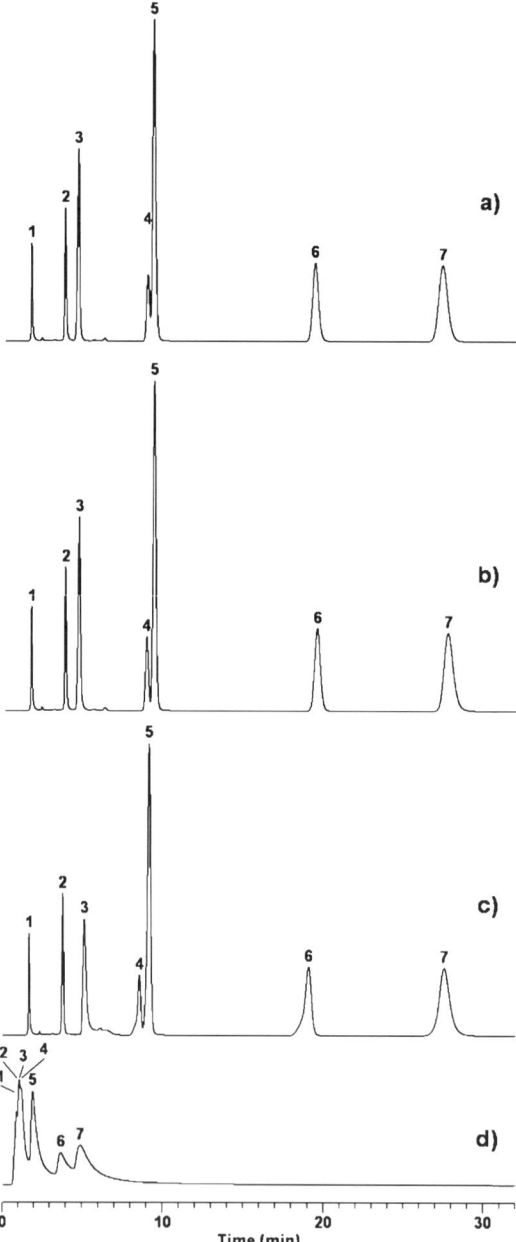

Figure 5.5 Temperature stability of a Waters XBridge column. Test chromatograms obtained (a) before the column was heated up to 150 °C; (b) after the fifth heating cycle; (c) after the tenth heating cycle; and (d) after the fifteenth heating cycle. Chromatographic conditions: mobile phase: phosphate buffer of pH 7–methanol (35 : 65 v/v); flow rate: 1 ml min^{-1}; detection: UV at 254 nm; temp.: 25 °C. Peaks: 1, dihydroxyacetone; 2, propyl paraben; 3, propranolol; 4, dipropyl phthalate; 5, naphthalene; 6, acenaphthene; and 7, amitriptyline.[10] (Reproduced with kind permission from Wiley-VCH Verlag GmbH & Co. KGaA.)

stationary phases, a new "boost" in high-temperature HPLC has begun. Please note that I will not give the full experimental evidence for all silica-based columns we have tested in this monograph. Here, I advise the reader to consult the respective literature. However, I have summarized these findings in Table 5.1 at the end of this chapter and have made suggestions regarding the highest temperatures at which these columns should be used. I would also like to emphasize that this table is not comprehensive, as there are other materials which might be equally stable. Therefore, I would be more than happy for column manufacturers to donate columns to our laboratory for us to test them under the conditions described above.

Nevertheless, alternative materials based on either zirconium or titanium dioxide will be indispensable when new hyphenation techniques are concerned, as I will demonstrate in Chapters 6 and 8. In the following section I will therefore provide some information about the stability of these stationary phases which have been used over a long period in high-temperature HPLC.

5.4 Zirconium Dioxide Stationary Phases

Stationary phases based on zirconium dioxide were introduced by Carr and co-workers some 20 years ago.[25] The solubility of bare zirconium dioxide in hot water is undoubtedly much less than that of bare silica. Therefore, zirconium dioxide appears to be an ideal support for high-temperature HPLC. However, until now, there has been no comparable mechanism of bonding a silane to the metal oxide surface. Hence, the surface modification is achieved by covering the surface with a coating of polymer or a cladding of carbon.[25–28] The most popular surface modification is polybutadiene (PBD) which covers the metal oxide.[29–45] For purely hydrophobic compounds, the retention mechanism is comparable to silica-based reversed-phase materials.[32] However, if polar compounds with ionisable groups are involved, a different selectivity is obtained when compared to that of silica-based phases.[5,37,41,46] This is because the zirconium dioxide can interact with polar analytes *via* Lewis acid–base interactions. Although the selectivity is one of the most important parameters in order to enhance the resolution of critical peak pairs, there is still a widespread reluctance to use these phases in the pharmaceutical industry. This is really a pity because these phases cannot only be used at elevated temperatures, but also with highly acidic or basic pH eluents. Therefore, the elution of basic substances could be greatly improved, but large-scale studies have not been conducted in industry to explore the full potential of these materials. The interested reader is referred to two very good reviews which explain the multiple interactions between the solute and the stationary phase.[3,4]

When it comes to special hyphenation techniques which require the use of a pure water mobile phases, we have had very good experiences with these columns in our laboratory. Indeed, there is no need to worry about phase de-wetting or a phase collapse if only water is used as the mobile phase over a prolonged period. Figure 5.6 depicts four chromatograms which have been collected after the column was exposed to a mobile phase of water–methanol (90 : 10 v/v) at 150 °C at neutral (see Figure 5.6b), acidic (see Figure 5.6c) and basic (see Figure 5.6d) conditions for 25 hours each. I would like to highlight two important points when this data is compared to the chromatograms given in the previous section (*e.g.* the ethylene-bridged hybrid silica-based Waters XBridge column). Firstly it appears that after the neutral elution phase, there is a slight decrease in the retention factors of test solutes. This means that the stationary phase becomes more polar, revealing that the polymer is slowly washed off the column.

This is consistent with experiments I have described in Chapter 5.1 to measure the column bleed under temperature-programmed conditions (see Figures 5.1 and 5.2).[8] Nevertheless, a high efficiency is maintained after the neutral heating phase,because the peak asymmetry and the plate number are hardly affected. When the pH is now adjusted to 2.2 using phosphate buffer a severe tailing and even peak splitting can be noticed (see Figure 5.6c). The column was then regenerated according to the procedure recommended by the column manufacturer, but we were not able to fully restore its initial efficiency. After the basic heating phase at pH 12, there is no complete loss in retention as experienced by the silica-based stationary phases, but it is obvious that the retention of the polar analytes is severely affected. This is because a phosphate buffer can induce a positive or negative charge on the surface, depending on the pH of the mobile phase. However, a direct comparison with the XBridge phase reveals that the overall stability of the PBD-coated zirconium dioxide phase at high temperature and basic pH is unparalleled.

Although this appears to be good news, I have to point out that these columns are prone to degradation when they are used over a long period at temperatures above 150 °C. We have used a lot of these columns during the last few years, and it appears that the column manufacturer has significantly improved the coating process. In the past, many authors have shown that the overall efficiency of metal oxide supports appears to be much lower than that for silica-based phases. However, this assumption is no longer valid since the new generation of these phases now has a comparable efficiency to silica-based phases, this being recently confirmed by Silva *et al.*[47] Unfortunately, improving the coating procedure, which means that the thickness of the film is thinner or the polymer is distributed more evenly in the pores, leads to a shorter lifetime of these columns at elevated temperatures. So while the performance has been improved for "ambient" temperature separations, the ruggedness is now reduced for high-temperature applications.

Another point I would like to highlight is that polymer-coated metal oxide stationary phases exhibit a lower hydrophobicity than ODS modified silica-based packings. Therefore, much less organic solvent is required with these columns to achieve the same retention of hydrophobic compounds. This is a decisive advantage when hyphenation techniques are applied which are based on the use of a pure water mobile phase. Indeed, the complete reduction of the

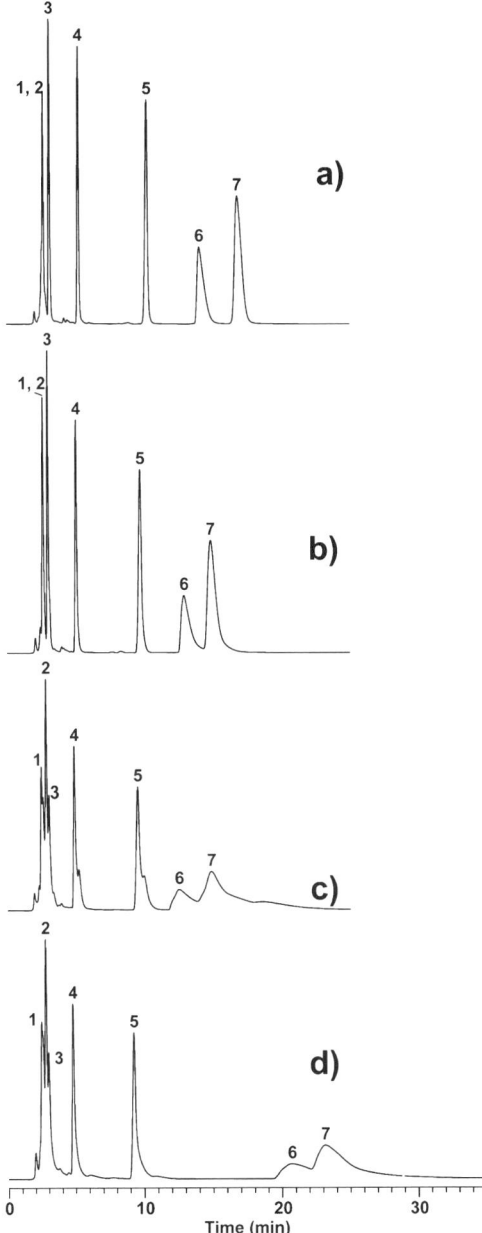

Figure 5.6 Temperature stability of a ZirChrom-PBD column. Test chromatograms obtained (a) before the column was heated up to 150 °C; (b) after the fifth heating cycle; (c) after the tenth heating cycle; and (d) after the fifteenth heating cycle. Chromatographic conditions: mobile phase: phosphate buffer of pH 7–methanol (35 : 65 v/v); flow rate: 0.7 ml min^{-1}; detection: UV at 254 nm; temp.: 25 °C. Peaks: 1, dihydroxyacetone; 2, propyl paraben; 3, dipropyl phthalate; 4, naphthalene; 5, acenaphthene; 6, propranolol; and 7, amitriptyline.[10] (Reproduced with kind permission from Wiley-VCH Verlag GmbH & Co. KGaA.)

organic co-solvent is much easier to achieve on polymer-coated packings than on silica-based phases. This is very relevant for some special hyphenation techniques described in Chapter 8. It is then essential to find a suitable stationary phase with a low hydrophobicity in order to enable the elution of polar and non-polar compounds in the same chromatographic run. *In this respect, polymer-coated zirconium dioxide stationary phases have a clear advantage, when compared to ODS reversed-phase materials, that an elution of these compounds can be achieved with either a lower concentration of the organic co-solvent or at a lower temperature under otherwise identical conditions.*

Therefore, I can strongly recommend these kinds of phases for all applications where a pure water mobile phase needs to be used. I will show some applications in Chapters 6 and 8 which support this conclusion.

The coating of zirconium dioxide stationary phases is not limited to polybutadiene, since polystyrene-coated zirconium dioxide, known as ZirChrom-PS, is also commercially available. However, it appears that PBD-coated zirconium dioxide is by far the most widely used. Besides the coating process, the cladding of carbon onto the surface is also possible in order to obtain either carbon-cladded zirconium dioxide or a modification where C-18 is covalently attached to the carbon-cladded surface. These phases are known as ZirChrom-Carb and ZirChrom-Diamond Bond, which are more hydrophobic than the PBD or PS-coated phases.

I would like to close this section with a description of the superior stability of a carbon-clad zirconia phase. This column was used in our laboratory for a long-term study of the phase's stability at extremely high temperature. It was indeed used more than 200 hours at a temperature of 185 °C, using a mobile phase of water and acetonitrile, each spiked with 0.1% formic acid. After this, the column was used for more than 100 hours at a maximum temperature of 225 °C while the pH was adjusted to 2 and 12 using phosphate buffer. We were quite astonished that this column produced good peak shapes after this harsh treatment, although the overall efficiency was very low. So whenever a separation has to be performed at extreme conditions, the use of carbon-cladded zirconium dioxide should be considered. For example, a carbon-clad zirconium dioxide column has also proven useful in a very special application of high-temperature HPLC. In a recent article about on-line two dimensional HPLC, Carr and co-workers used a system where they employed a carbon-clad zirconium dioxide stationary phase as the second dimension column.[48] For this method, the high retentivity of the carbon-clad stationary phase was very beneficial, because in this case, a focusing of analytes could be achieved even if the solvent delivered from the first dimension column has a high elution strength. In order to reduce the analysis time of the second dimension run, the column was operated at a high temperature and high flow rate. Now it is clear why a highly retentive column is the best choice for this kind of application. By increasing the temperature, the elution of the mobile phase is stronger when a solvent gradient is run at high temperature than at ambient temperature. Hence, the retentivity of the stationary phase should be very high, otherwise, the compounds would elute close to the dead time. The second advantage of employing high temperatures in on-line two dimensional HPLC is that the column can be operated at a high flow rate without a significant loss of efficiency. *In this respect, a carbon-clad zirconium dioxide column has the decisive advantage that a good retentivity can be maintained at high eluent temperatures. Therefore, this type of column is an ideal candidate for on-line two-dimensional liquid chromatography, where high thermal stability and high hydrophobicity are advantageous.*

Although I mentioned that these types of columns are not currently used as a first choice for most separation problems, I really hope that the attitude towards these phases will change. What should also be noted is that the hydrophobicity which is needed for a certain application or separation problem can be easily tuned with zirconia-based phases. If a method or technique requires a phase with a low hydrophobicity, a polymer-coated stationary phase should be used, while a carbon-clad phase is suitable for very polar analytes which are difficult to retain on any silica-based ODS column. Separation scientists often lament that the selectivities of most silica-based RP phases are not very different. Therefore, the best approach to resolve a critical peak pair which completely co-elutes on one column is to change the selectivity of the phase system. One option I described in Chapter 4 is to screen different mobile phases (see Figure 4.16). The other option would be to use a stationary phase with a different selectivity. My advice therefore is that you should try a column with a different selectivity in case you have a very difficult separation problem where compounds are completely co-eluting. Simply increasing the plate number will not always solve such a problem. In addition, temperature is a very powerful tool to change the selectivity as will be demonstrated in Chapter 6.

5.5 Titanium Dioxide Stationary Phases

Stationary phases based on titanium dioxide are probably even less well known than their zirconium dioxide analogues. This might be because metal oxide stationary phases are generally considered very exotic when compared to traditional silica-based columns. As with zirconium dioxide, the support can be coated with different polymers. One of these surface modifications consists of polyethylene. Besides a polymeric coating, cladding with carbon also leads to a reversed-phase material and the hydrophobicity of a carbon-cladded titanium dioxide surface is higher than the coating with polyethylene.

The chemical stability of titanium-based stationary phases is comparable with zirconium-based stationary phases. It was pure curiosity that led us to think about a very unusual experiment. During a conference someone came to me and asked if I had some information about the stability of a bare silica stationary phase, which means a normal phase. I said that we didn't test such a phase, but that it would be a very interesting experiment. So the aim of the experiment was to directly compare a bare silica stationary phase with a bare titanium dioxide stationary phase. The test procedure was the same which I outlined in section 5.2, except that we adjusted the maximum temperature to

120 °C for the silica and to 185 °C for the titanium dioxide stationary phase using pure water as the purging eluent. The results of this quite unusual experiment are depicted in Figures 5.7 and 5.8.

It can be clearly seen that the silica column was hydrolytically attacked and has totally collapsed after 30 hours. In contrast to this, hydrolysis of the titanium support material was not observed even though the temperature was

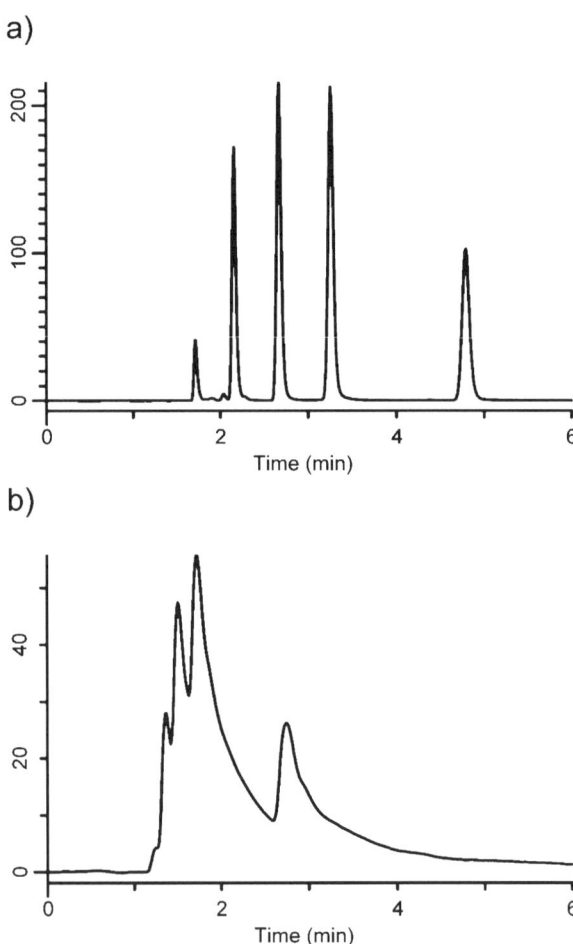

Figure 5.7 Test chromatograms using a Waters Spherisorb Silica column obtained (a) before the first and (b) after the sixth heating interval. Test solutes: toluene, dibutyl phthalate, diethyl phthalate, dimethyl phthalate and cinnamyl alcohol. Test solutes were dissolved in heptane–isopropanol (96 : 4). The test mixture was diluted with heptane–isopropanol (96 : 4 v/v). Chromatographic conditions: mobile phase flow rate: 1 ml min^{-1}; eluent temp.: 30 °C; mobile phase: heptane–isopropanol (96 : 4); detection: UV at 254 nm.[9] (Reproduced with kind permission from Wiley-VCH Verlag GmbH & Co. KGaA.)

Suitable Stationary Phases 103

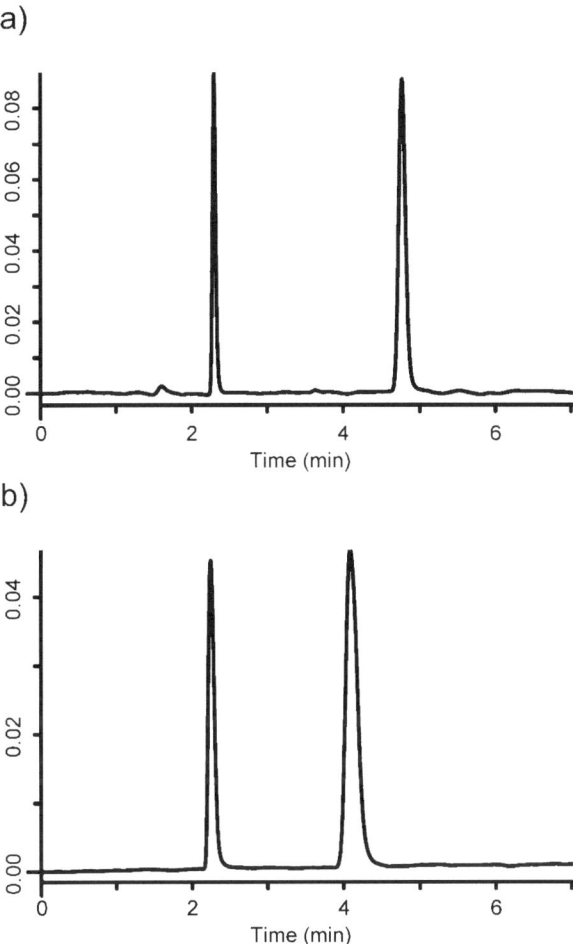

Figure 5.8 Test chromatograms using a Sachtopore NP column obtained (a) before the first and (b) after the tenth heating interval. Test solutes: benzene and nitrobenzene. Test solutes were dissolved in *n*-heptane. The test mixture was diluted with *n*-heptane. Chromatographic conditions: flow rate: 1 ml min^{-1}; eluent temp.: 30 °C; mobile phase: *n*-heptane; detection: UV at 254 nm.[9] (Reproduced with kind permission from Wiley-VCH Verlag GmbH & Co. KGaA.)

much higher and the test procedure was stopped after 50 hours. However, it was noticed that the peaks had broadened and that there was a slight shift in the retention of test solutes. It can be assumed that such a peak broadening results if the column is subjected to thermal stress as was outlined (see my comment on page 89). Repeated heating up and cooling down of the column might lead to a rearranging of the particles. A void volume can build up at the top or the wall of the column, reducing its efficiency and leading to broader peaks. We have often observed such a phenomenon where the initial performance of the column was immediately affected after just one heating cycle, but did not change further during the progression of the test procedure.

Another column based on titanium dioxide that we have used for many years is a reversed-phase stationary phase with a polyethylene coating. I'm not able to establish the exact length of time that the column was used at a certain temperature, but one of the two columns was used over a period of about four years and was still not fully "destroyed" by the high temperatures. However, there are two problems I have to address. While a column with a polyethylene coating is ideally suited for a pure water mobile phase, a sudden and rapid stripping of the polymer is noticed when a binary mobile phase of water and an organic co-solvent is used. In this case, the polymer is dissolved and washed off the column. We noticed this phenomenon when working with a mobile phase of water and tetrahydrofuran at elevated temperatures. Suddenly, the noise of the UV detector increased tremendously which was due to a sudden stripping off of the polymer. When the column was cooled down and a test mixture was measured at ambient temperature, a dramatic loss in the retentivity of the column was observed. You could probably use this procedure to transform the reversed-phase to a normal-phase column. Just add an organic co-solvent to the mobile phase, make it hot and purge the column for 10 hours. I guess that the polymer should be completely removed and you end with a bare titanium dioxide phase. However, this was not our intention and therefore I recommend that this column is only used with a pure water mobile phase at high temperatures. For all applications which require a water-only mobile phase, a polyethylene-coated titanium dioxide stationary phase is ideally suited and very robust. We couldn't even notice a loss in retention once the column had been heated up to 185 °C for some months. However, I must also admit that the efficiency of the column was not very satisfactorily. I think that this can be attributed to the fact that the packing procedure needs to be optimized. As I already stressed, the formation of voids and channels through the packed bed will result in additional band broadening, reducing the efficiency of the column after a few heating cycles.

It can be summarized that titanium-based stationary phases are ideally suited for high-temperature HPLC. But you have to be careful when using a reversed-phase column with a water–organic mobile phase. Be sure to completely remove the organic portion in the eluent before you start to heat up the column. Otherwise, you will notice a rise in the baseline, which can be attributed to the stripping off of the polymer. For water-only separations, however, this column is ideally suited and you need never worry about a phase collapse due to dewetting.

5.6 Polymeric Stationary Phases

Traditionally, polymeric phases have been used in gel-permeation chromatography (GPC) or size-exclusion chromatography (SEC). The difference between these techniques is that in GPC only pure organic mobile phases are used, while in SEC water is used as the sole eluent. The separation mechanism is not usually based on an interaction of the solutes with the stationary phase. In GPC as well as SEC, the separation is obtained by molecular size, which often corresponds to molecular weight. The mobile phase is largely "passive" as is the carrier gas in GC, with the only function to transport the solutes through the column and to make sure that the solutes, which are mainly polymers, do not precipitate. In GPC, THF is mostly used as the sole eluent because it has good solubilizing properties for many polymers. However, GPC is often carried out at very high temperatures. Therefore, people who are involved in GPC analyses are used to applying temperatures of up to 150 °C or even higher. Moreover, the mobile phases are often quite "nasty", with a high potentially toxicity.

Unsurprisingly, many separations in high-temperature liquid chromatography have been carried out on stationary phases which are suitable for GPC analyses. Mostly, a polystyrene–divinylbenzene (PS–DVB) packing has been employed.[7,49–63] Although these phases are very rugged towards high eluent temperatures and extremely low and high pH, a traditional weakness is their low mechanical stability and low efficiency. In contrast to this, the enormous success of silica-based packings is based on their high mechanical strength combined with a very high efficiency. If you compare a separation on a silica-based ODS phase with a separation on a PS–DVB packing, you will be disappointed, because on the GPC column, peaks will be very broad. Furthermore, such phases can be extremely hydrophobic, which means that a higher content of organic modifier is required to achieve the same retention when compared to a silica-based RP phase.

Another problem which is inherent with these packings is that they may swell or shrink as the mobile phase composition is changed. Usually, in GPC or SEC analyses, you will either use pure organic solvents or pure water. Hence, the polymeric network is adapted to the solvent. In case of reversed-phase HPLC, solvent-gradient programming is applied, which means that you start with a high content of water and gradually increase the organic portion in the mobile phase. This can lead to a swelling or shrinking of the polymeric network. In this case, we have often observed that the pressure will increase when a GPC column is used in solvent-gradient mode at elevated temperature. Such an increase in pressure can trigger the collapse of the phase, because once the column is operated at the maximum pressure, it can collapse so that the pores are blocked and the flow must be stopped. However, other groups reported about a successful implementation of polymeric phases in high-temperature liquid chromatography.

It can be summarized that the use of polymeric phases for high-temperature liquid chromatography is not trouble-free, although they are quite stable at high temperature. Unfortunately, they often suffer from a slow mass transfer, resulting in a lower efficiency than silica-based phases. Also, the polymeric network can swell or shrink if a solvent gradient is applied.

5.7 Other Materials

5.7.1 Graphitized Carbon Column

There are also other materials which can be used at high eluent temperatures. One of these materials is comprised of pure graphitized carbon (PGC). This column, which is also known under its trade name Hypercarb®, has a completely different retention mechanism and can be used for the separation of structurally similar compounds like stereoisomers. I would advise the reader to consult two exhaustive articles published by Knox and Ross, where the retention mechanism as well as some applications are described.[64,65] The fact that this column consists of pure carbon means that it is virtually unaffected by high eluent temperatures. However, some years ago, the columns suffered from a rapid breakdown at elevated temperatures and again it appeared that the marketing division was one step ahead of the research division, so I have to tell another anecdote. When the interest in high-temperature HPLC was renewed, the manufacturer discovered that he had a material with an unlimited temperature and pH stability. So the marketing division advertised that this column was ultimately robust towards high eluent temperatures. When I worked on my PhD, I eagerly noticed the advertisement and used such a column. After the first heating cycle I noticed that there was a leakage when the column was cooled down, so I had to retighten the fittings and hoped that the problem would be solved. Unfortunately, all my endeavours failed, so I contacted the manufacturer directly and explained the situation. After a while it was discovered that the O-rings which should seal the column were made of polyetheretherketone (PEEK) and thus, a leakage was inevitable. Now, the complete column hardware is available in stainless steel and can be used at high temperatures without the fear of mobile phase leakage.[ii] This story is typical for the progress which is made when a new technique gradually emerges. At the first stage, the marketing department highlights the general advantages of a column or packing material. Due to the fact that only a few customers use the new technique, there is limited experience regarding the fidelity of these specifications. In the next stage, the deficiencies are corrected and a new hardware is available. The problem is that the notions regarding such difficulties are long-standing and much effort is needed to convince customers that the operation of the new hardware is trouble-free. Therefore, column manufacturers need to apply test procedures before a new product is launched in order to guarantee that the specifications of their materials will be met. Otherwise, the damage is higher than the benefit. *This example also highlights that it is not only the stationary phase itself which must be thermally stable, but also the hardware of the column has to be considered.*

As I mentioned, the retention and selectivity of a PGC column is completely different from those of silica-based phases. The mechanism has been explained by an effect which is called the "Polar Retention Effect on Graphite" (PREG). This means that the more polar the compounds in a homologues series, the more retained the compounds will be.[64,65] Therefore, PGC is particularly useful for the separation of highly polar compounds which would be difficult to retain on an ODS phase.

[ii] Please note that the last numbers of the column's part number should end with "46". Then you can be sure that the complete column hardware is made up of stainless steel. If the part number ends with "30", PEEK has been used for the O-rings.

I must add, however, that the hydrophobic surface strongly retains non-polar compounds. In many cases it is extremely difficult to elute these compounds even with a very strong solvent. This can lead to unwanted bleed over a long period. Nevertheless, if highly polar compounds have to be separated which cannot be retained on any other material, a Hypercarb column might solve the problem. However, care should be taken if the analyte is contained in a complex matrix, because then the matrix might be irreversibly adsorbed onto the stationary phase. In fact, a regeneration of the column can be quite difficult when matrix compounds are adsorbed onto the surface. This will result in a slow but irreversible contamination of the complete column.

Although this may sound a little bit too negative, these phases have been successfully employed for a range of high-temperature separations. The interested reader is referred to the respective literature.[48,66–69]

5.7.2 Thermo-Responsive Stationary Phases

I will close this section with a very interesting approach to column chemistry, which is also overlooked by the chromatographic community in general. As was already mentioned, retention usually decreases when the temperature is increased.[iii] However, a concept which is known as "thermo-responsive chromatography" uses the adverse effect, which means that retention can be increased with increasing temperature. These kinds of stationary phases are also based on silica gel, which has been modified with temperature-responsive polymers. Here, poly(N-isopropylacrylamide) is grafted onto (aminopropyl)silica using an activated ester–amine coupling method. These grafted silica surfaces then exhibit hydrophilic properties at lower temperatures, which transform to hydrophobic surface properties when the temperature is increased. This work has been carried out mainly by Kanazawa and co-workers.[70–81] The effect is based on a phase transition of aqueous poly(N-isopropylacrylamide) and occurs around a lower critical solution temperature of about 32 °C. This phase transition is reversible and also reproducible, which is a prerequisite for its usage in liquid chromatography. It arises due to conformational changes of the polymer chain. The synthesis and application of these phases have been published in a number of articles. Since the interested reader will find more details in these references, I will only discuss a few examples and then highlight why such phases might also be successfully used in high-temperature HPLC.

Figure 5.9 shows a separation of a test mixture containing five steroids with a pure water mobile phase. It can be clearly seen that with increasing temperature, retention is also increasing. Furthermore, the elution of these compounds can be achieved with a pure water mobile phase without the need of an organic modifier. This is not only favourable in saving high amounts of toxic

---

[iii] I will explain the theoretical back-ground of the temperature dependence of analyte retention in Chapter 6.4.1. For the moment it is sufficient to know that retention normally decreases with increasing temperature.

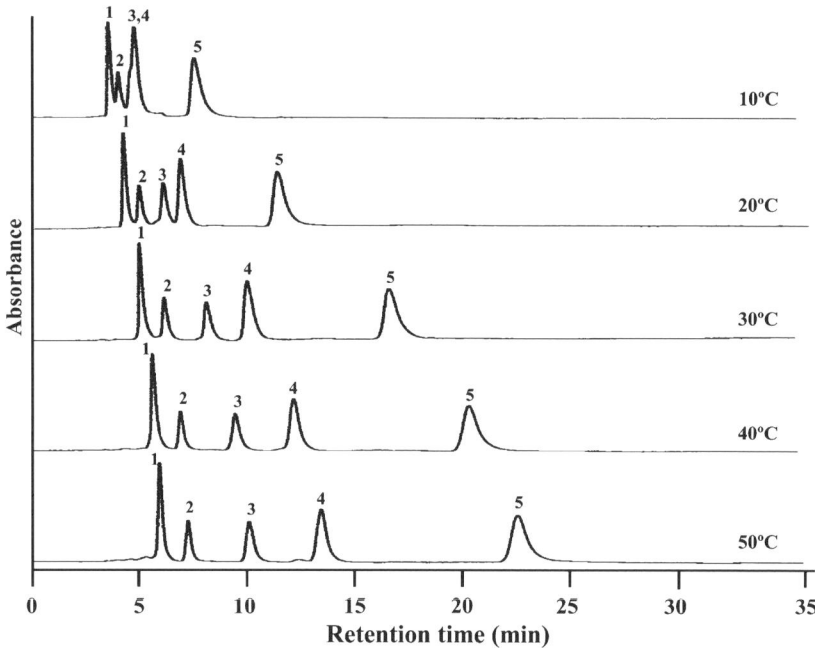

Figure 5.9 Chromatograms of a mixture of five steroids with water as mobile phase at various temperatures on a PNIPAAm-modified column (4.6 × 150 mm ID). Peaks: 1, hydrocortisone; 2, prednisolone; 3, dexamethasone; 4, hydrocortisone acetate; 5, testosterone. Chromatographic conditions: mobile phase: water; flow rate: 1.0 ml min^{-1}; detection: UV at 254 nm.[80] (Reproduced with kind permission from Wiley-VCH Verlag GmbH & Co. KGaA.)

and costly solvents, but also the hyphenation with the special detection systems that I will discuss in Chapter 8 becomes feasible. Since steroids are relatively non-polar compounds, the content of the organic solvent often has to be increased significantly in order to achieve an elution of these substances on a traditional silica-based reversed-phase stationary phase.

The analysis of the retention mechanism of these kinds of stationary phase materials reveals that there seem to be at least two regions in which a linear van't Hoff plot can be assumed. Figure 5.10 depicts the van't Hoff plot of the separated compounds shown in Figure 5.9. Obviously, the slope of this plot is negative for all compounds, which is completely different to a classical reversed-phase separation mechanism. Also, it can be noticed that around the lower critical solution temperature, the phase transition of the stationary phase is responsible for the non-linearity of the retention mechanism.

Since the temperature in most studies has not been increased much beyond 50 °C, it does not seem appropriate to speak of high-temperature HPLC when you remember the definition I gave in Chapter 1. However, it would be very

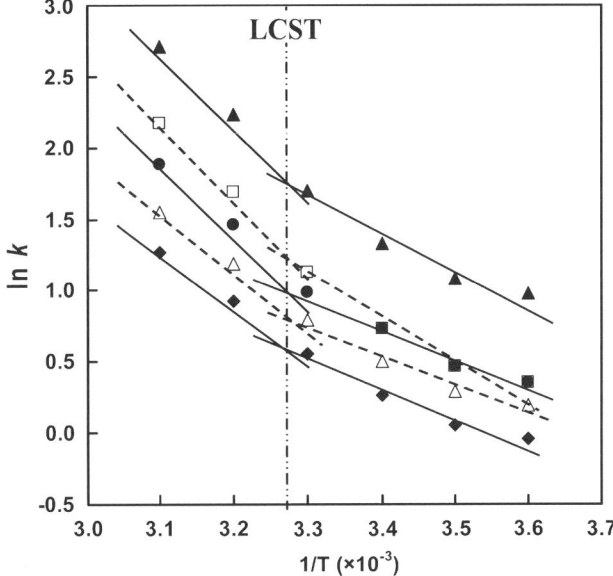

Figure 5.10 Van't Hoff plots of steroids with water as mobile phase on PNIPAAm-modified column (4.6 × 150 mm ID). ◆, Hydrocortisone; △, prednisolone; ●, dexamethasone; □, hydrocortisone acetate; and ▲, testosterone. Chromatographic conditions: mobile phase: water; flow rate: 1.0 ml min^{-1}; detection: UV at 254 nm.[80] (Reproduced with kind permission from Wiley-VCH Verlag GmbH & Co. KGaA.)

interesting to increase the lower critical solution temperature further or even to modify the polymer chains so that a multiple phase transition can be achieved. In this case, negative temperature programming could be used to replace solvent programming for many compounds. What needs to be considered is that the heating system also has to be modified. While systems for temperature programming in the positive mode are readily available, homemade systems have to be used for negative temperature programming.

5.8 General Conclusions

As I have already mentioned, the stability of the column at high eluent temperatures can be considered a major task which needs to be further improved by column manufacturers. The results I have given in this chapter showing the stability at very high temperatures and extreme pH might look discouraging. However, it should be kept in mind that the test procedure I have presented was aimed to facilitate a rapid breakdown of the selected columns. By increasing the temperature to 150 °C we were able to clearly differentiate between columns which can be really used in high-temperature HPLC at temperatures above

Table 5.1 Overview of the stationary phases which can be used at elevated temperatures.

Manufacturer	Column description	Maximum temperature/°C
Agilent	Zorbax SB C-18	100
Phenomenex	Gemini NX C18	120
Polymer Laboratories	PLRP-S	200
Restek	pHidelity C-18	100
Sachtleben	Sachtopore NP	200[a]
Sachtleben	Sachtopore RP	150[b]
Selerity	Blaze 200 C-18	100
Thermo	Hypercarb	200
Waters	XBridge C-18	150–180
ZirChrom	Carb	200
ZirChrom	Phase	200[a]
ZirChrom	PBD	120–150
ZirChrom	PS	120–150

[a]This is a normal phase column without a surface modification.
[b]This phase should only be used with a pure water mobile phase at temperatures above 80 °C.

100 °C. This does not mean that all other columns which failed are not suitable for high-temperature HPLC applications, but that the maximum temperature for long-term operation is lower. In Table 5.1 I have listed all columns we have tested in recent years according to the procedure I have given in section 5.2, and which I would recommend for use in high-temperature HPLC. When the upper temperature limit is approached, the pH of the mobile phase should lie between 3 and 6. A higher or lower pH might lead to a faster degradation and thus, the maximum temperature should be decreased to prolong the column lifetime. Please note that there may be other columns which also have an enhanced temperature stability. I think that column manufacturers will continue to improve the stability of their materials over the years to come.

When the question is simply to increase the speed of a separation, a temperature around 100 °C will in most cases lead to a significant acceleration of the elution as was shown in Chapter 4. In this respect, other materials like the Zorbax StableBond column, the pHidelity or Blaze column can be used. In contrast to this, a higher stability is needed when special hyphenation techniques are employed and the mobile phase consists of pure water. Therefore, the development of new materials which can be specifically used under such harsh conditions is not over, but will continue into the foreseeable future. But this should not prevent us from abandoning this very promising technique. The commercial availability of hybrid silica particles with an extended pressure, temperature and pH range is really a boost for high-temperature HPLC, because silica-based packings are clearly the "gold-standard" in the pharmaceutical industry. Nevertheless, I'm also quite optimistic that alternative materials will be used on a large scale once the methods and techniques I will describe in Chapter 8 become standard procedures in routine laboratories.

References

1. U. D. Neue, *HPLC Columns*, Wiley-VCH, Weinheim, 1997.
2. J. Nawrocki, M. P. Rigney, A. McCormick and P. W. Carr, *J. Chromatogr., A*, 1993, **657**, 229.
3. J. Nawrocki, C. Dunlap, A. McCormick and P. W. Carr, *J. Chromatogr., A*, 2004, **1028**, 1.
4. J. Nawrocki, C. Dunlap, J. Li, J. Zhao, C. V. McNeff, A. McCormick and P. W. Carr, *J. Chromatogr., A*, 2004, **1028**, 31.
5. J. Dai, X. Yang and P. W. Carr, *J. Chromatogr., A*, 2003, **1005**, 63.
6. D. A. Fonseca, H. R. Gutierrez, K. E. Collins and C. H. Collins, *J. Chromatogr., A*, 2004, **1030**, 149.
7. I. D. Wilson, *Chromatographia*, 2000, **52**, 28.
8. T. Teutenberg, J. Tuerk, M. Holzhauser and T. K. Kiffmeyer, *J. Chromatogr., A*, 2006, **1119**, 197.
9. T. Teutenberg, J. Tuerk, M. Holzhauser and S. Giegold, *J. Sep. Sci.*, 2007, **30**, 1101.
10. T. Teutenberg, K. Hollebekkers, S. Wiese and A. Boergers, *J. Sep. Sci.*, 2009, **32**, 1262.
11. P. He and Y. Yang, *J. Chromatogr., A*, 2003, **989**, 55.
12. Y. Yang, *LCGC Eur.*, 2003, **16**, 37.
13. H. A. Claessens, M. A. van Straten and J. J. Kirkland, *J. Chromatogr., A*, 1996, **728**, 259.
14. G. Guiochon, *J. Chromatogr., A*, 2006, **1126**, 6.
15. U. D. Neue, E. Serowik, P. Iraneta, B. A. Alden and T. H. Walter, *J. Chromatogr., A*, 1999, **849**, 87.
16. M. Przybyciel and R. E. Majors, *LCGC Eur.*, 2002, **15**, 652.
17. S. Heinisch, G. Desmet, D. Clicq and J. L. Rocca, *J. Chromatogr., A*, 2008, **1203**, 124.
18. H. G. Gika, G. Theodoridis, J. Extance, A. M. Edge and I. D. Wilson, *J. Chromatogr., B: Biomed. Appl.*, 2008, **871**, 279.
19. R. Plumb, J. R. Mazzeo, E. S. Grumbach, P. Rainville, M. Jones, T. Wheat, U. D. Neue, B. Smith and K. A. Johnson, *J. Sep. Sci.*, 2007, **30**, 1158.
20. D. T. T. Nguyen, D. Guillarme, S. Heinisch, M. P. Barrioulet, J. L. Rocca, S. Rudaz and J. L. Veuthey, *J. Chromatogr. A*, 2007, **1167**, 76.
21. S. Shen, H. Lee, J. McCaffrey, N. Yee, C. Senanayake and N. Grinberg, *J. Liq. Chromatogr. Relat. Technol.*, 2006, **29**, 2823.
22. Y. Liu, N. Grinberg, K. C. Thompson, R. M. Wenslow, U. D. Neue, D. Morrison, T. H. Walter, J. E. O. Gara and K. D. Wyndham, *Anal. Chim. Acta*, 2005, **554**, 144.
23. L. Al-Khateeb and R. M. Smith, *J. Chromatogr., A*, 2008, **1201**, 61.
24. L. A. Al-Khateeb and R. M. Smith, *Anal. Bioanal. Chem.*, 2009, **394**, 1255.
25. M. P. Rigney, T. P. Weber and P. W. Carr, *J. Chromatogr.*, 1989, **484**, 273.

26. T. P. Weber, P. T. Jackson and P. W. Carr, *Anal. Chem.*, 1995, **67**, 3042.
27. J. Zhao and P. W. Carr, *Anal. Chem.*, 1999, **71**, 5217.
28. E. F. Funkenbusch, P. W. Carr, D. A. Hanggi and T. P. Weber, *Carbon-Clad Zirconium Oxide Particles, Patent US 5346619*, University of Minnesota, Minneapolis, MN, 1994.
29. L. F. Sun, A. V. Mccormick and P. W. Carr, *J. Chromatogr., A*, 1994, **658**, 465.
30. L. F. Sun and P. W. Carr, *Anal. Chem.*, 1995, **67**, 2517.
31. L. F. Sun and P. W. Carr, *Anal. Chem.*, 1995, **67**, 3717.
32. J. Li and P. W. Carr, *Anal. Chem.*, 1996, **68**, 2857.
33. J. Li and P. W. Carr, *Anal. Chem.*, 1997, **69**, 2202.
34. J. Li and P. W. Carr, *Anal. Chem.*, 1997, **69**, 2193.
35. J. W. Li, Y. Hu and P. W. Carr, *Anal. Chem.*, 1997, **69**, 3884.
36. Z.-T. Jiang, D.-Y. Zhang and Y.-M. Zuo, *J. Liq. Chromatogr. Relat. Technol.*, 2000, **23**, 1159.
37. Y. Hu, X. Q. Yang and P. W. Carr, *J. Chromatogr., A*, 2002, **968**, 17.
38. T. Andersen, Q. N. Nguyen, R. Trones and T. Greibrokk, *J. Chromatogr., A*, 2003, **1018**, 7.
39. Y. Xiang, B. Yan, B. Yue, C. V. McNeff, P. W. Carr and M. L. Lee, *J. Chromatogr., A*, 2003, **983**, 83.
40. M. M. Sanagi, H. H. See, W. A. Ibrahim and A. A. Naim, *J. Chromatogr., A*, 2004, **1059**, 95.
41. C. Ovens, D. Sievwright and A. J. Silcock, *J. Chromatogr., A*, 2006, **1137**, 56.
42. S. M. Fields, C. Q. Ye, D. D. Zhang, B. R. Branch, X. J. Zhang and N. Okafo, *J. Chromatogr., A*, 2001, **913**, 197.
43. T. S. Kephart and P. K. Dasgupta, *Talanta*, 2002, **56**, 977.
44. O. Chienthavorn, R. M. Smith, I. D. Wilson, B. Wright and E. M. Lenz, *Phytochem. Anal.*, 2005, **16**, 217.
45. T. S. Kephart and P. K. Dasgupta, *Anal. Chim. Acta*, 2000, **414**, 71.
46. X. Q. Yang, J. Dai and P. W. Carr, *J. Chromatogr., A*, 2003, **996**, 13.
47. C. R. Silva, C. H. Collins, K. E. Collins and C. Airoldi, *J. Sep. Sci.*, 2006, **29**, 790.
48. D. R. Stoll, J. D. Cohen and P. W. Carr, *J. Chromatogr., A*, 2006, **1122**, 123.
49. D. J. Miller and S. B. Hawthorne, *Anal. Chem.*, 1997, **69**, 623.
50. Y. Yang, A. D. Jones and C. D. Eaton, *Anal. Chem.*, 1999, **71**, 3808.
51. Y. Yang, A. D. Jones, J. A. Mathis and M. A. Francis, *J. Chromatogr., A*, 2002, **942**, 231.
52. Y. Yang, L. J. Lamm, P. He and T. Kondo, *J. Chromatogr. Sci.*, 2002, **40**, 107.
53. T. Kondo and Y. Yang, *Anal. Chim. Acta*, 2003, **494**, 157.
54. H. Chen and C. Horvath, *Anal. Meth. Instr.*, 1993, **1**, 213.
55. R. M. Smith and R. J. Burgess, *Anal. Commun.*, 1996, **33**, 327.
56. R. M. Smith and R. J. Burgess, *J. Chromatogr., A*, 1997, **785**, 49.

Suitable Stationary Phases

57. B. A. Ingelse, H.-G. Janssen and C. A. Cramers, *J. High Resolut. Chromatogr.*, 1998, **21**, 613.
58. O. Chienthavorn and R. M. Smith, *Chromatographia*, 1999, **50**, 485.
59. T. Teutenberg, O. Lerch, H.-J. Götze and P. Zinn, *Anal. Chem.*, 2001, **73**, 3896.
60. T. Yarita, R. Nakajima and M. Shibukawa, *Anal. Sci.*, 2003, **19**, 269.
61. O. Chienthavorn, R. M. Smith, S. Saha, I. D. Wilson, B. Wright, S. D. Taylor and E. M. Lenz, *J. Pharm. Biomed. Anal.*, 2004, **36**, 477.
62. T. Yarita, R. Nakajima, K. Shimada, S. Kinugasa and M. Shibukawa, *Anal. Sci.*, 2005, **21**, 1001.
63. M. M. Sanagi and H. H. See, *J. Liq. Chromatogr. Relat. Technol.*, 2005, **28**, 3065.
64. J. H. Knox and P. Ross, *Adv. Chromatogr.*, 1997, **37**, 73.
65. P. Ross and J. H. Knox, *Adv. Chromatogr.*, 1997, **37**, 121.
66. L. Pereira, *LCGC North Am.-Appl. Noteb.*, 2006, **June 2006**, 75.
67. L. Pereira, S. Aspey and H. Ritchie, *J. Sep. Sci.*, 2007, **30**, 1115.
68. D. Guillarme, S. Heinisch and J. L. Rocca, *J. Chromatogr., A*, 2004, **1052**, 39.
69. D. Guillarme, S. Heinisch, J. Y. Gauvrit, P. Lanteri and J. L. Rocca, *J. Chromatogr., A*, 2005, **1078**, 22.
70. H. Kanazawa, K. Yamamoto and Y. Matsushima, *Anal. Chem.*, 1996, **68**, 100.
71. H. Kanazawa, Y. Kashiwase, K. Yamamoto, Y. Matsushima, A. Kikuchi, Y. Sakurai and T. Okano, *Anal. Chem.*, 1997, **69**, 823.
72. H. Kanazawa and Y. Matsushima, *Yakugaku Zasshi*, 1997, **117**, 817.
73. H. Kanazawa, S. Tastuo and M. Yoshikazu, *Anal. Chem.*, 2000, **72**, 5961.
74. H. Kanazawa and Y. Matushima, *Anal. Sci.*, 2002, **18**, 45.
75. H. Kanazawa and Y. Matsushima, *Adv. Chromatogr.*, 2002, **41**, 311.
76. C. Sakamoto, Y. Okada, H. Kanazawa, E. Ayano, T. Nishimura, M. Ando, A. Kikuchi and T. Okano, *J. Chromatogr., A*, 2004, **1030**, 247.
77. H. Kanazawa, *Anal. Bioanal. Chem.*, 2004, **378**, 46.
78. E. Ayano, Y. Okada, C. Sakamoto, H. Kanazawa, T. Okano, M. Ando and T. Nishimura, *J. Chromatogr., A*, 2005, **1069**, 281.
79. H. Kanazawa, E. Ayano, C. Sakamoto, R. Yoda, A. Kikuchi and T. Okano, *J. Chromatogr., A*, 2006, **1106**, 152.
80. E. Ayano and H. Kanazawa, *J. Sep. Sci.*, 2006, **29**, 738.
81. E. Ayano, Y. Okada, C. Sakamoto, H. Kanazawa, A. Kikuchi and T. Okano, *J. Chromatogr., A*, 2006, **1119**, 51.

CHAPTER 6
Method Development using Temperature as an Active Variable

Many authors have stated that temperature can be regarded as the most underestimated parameter in liquid chromatographic separations, which had been noted already by Maggs in 1969.[1] Often, the positive contribution of temperature to solving difficult separation problems has been completely neglected, as was mentioned by Chmielowiec in 1979.[2] Many people have argued that temperature must not be changed because it has an effect on the retention of solutes and the selectivity of the separation. Well, this is really a terrific explanation why you should not play around with temperature!

Before I will start to explore the effects of temperature on resolution, including the effects on retention, selectivity and efficiency, I will shortly repeat and expand on some important facts which should be remembered.

6.1 Special Requirements of the Heating System

The requirements of a heating system, as addressed in Chapter 3, imply that a dedicated system should be modular so that the eluent temperature can be controlled independently from the column temperature. Furthermore, the system should allow for a post-column cooling of the effluent. Currently, there are heating ovens on the market which can be used not only for isothermal high-temperature operation but also for temperature programming.[3-5] The functional principles of these systems, however, differ significantly as was also outlined in Chapter 3.

Another question is "what should be regarded as the upper temperature limit that a heating oven should be able to reach?" In Chapter 5 it was shown that the stability of the stationary phase can be regarded as the limiting factor. Although most column ovens designed for high-temperature HPLC have an upper temperature limit of 200 °C, extending this limit further should not be that difficult to achieve. In contrast, designing ultra-stable stationary phases is much more demanding. Therefore, an upper temperature limit of about 200 °C for the oven was selected.

When the heating system is used solely for isothermal separations, the only requirement is that the temperature difference between the mobile phase entering the column and the stationary phase should be as low as possible. However, in order to compensate for the effects which are related to frictional heating (see Chapter 3.1.2), the preheating temperature should be controlled independently from the column temperature. Often, a slightly lower temperature of the eluent is beneficial when the separation is carried out at very high pressure. Also, a short preheating capillary of no more than 20 cm in length should be used in order to keep the extra-column volume as low as possible. In principle, forced-air ovens can be used once thermal equilibrium is established between the mobile and stationary phases.

In contrast, the requirements of a heating system to carry out temperature programming are more demanding. Here, it must be ensured that there is an efficient heat transfer to the column wall as well as in the radial direction of the packed bed. In my opinion, this is best achieved if contact heating is employed. In order to guarantee that there is a fast cycle time, the heating system should also allow for a rapid cool-down after the temperature gradient is finished. It makes no sense to develop a fast method without considering the time for gradient re-equilibration. In Chapter 3, I have given experimental evidence that even block-heating systems with a high thermal mass are ideally suited for temperature programming. Additionally, when a temperature gradient is applied, the temperature of the eluent as it leaves the column also changes with time. Depending on the detector type, it is necessary to cool down the eluent to a constant temperature. Therefore, it might be a requirement for the heating system to be equipped with an active temperature control to cool down the effluent before it is introduced into the detector. However, this clearly depends on the detection method and in some cases it might also be favourable not to cool down the eluent, especially if mass spectrometric detection is used, as I will show at the end of this chapter. Such a system which fulfils all the above mentioned requirements is now commercially available and has been developed in cooperation with Scientific Instruments Manufacturer (SIM GmbH).[3]

6.2 Special Requirements of the Column Hardware

As was pointed out in Chapter 5, it is not only the column packing which has to be stable under isothermal conditions. Regardless of whether you want to use temperature programming or just run your method isothermally, the column will be used at different temperatures as a method is developed. Therefore, the column is subjected to thermal stress, which means that the hardware may well expand as the temperature is increased and shrink as the temperature is decreased (see my comments on page 89). This process can lead to a rearrangement of particles and to the creation of voids, decreasing the initial efficiency of the column. Hence, the packing material not only needs to be robust against a thermally initiated hydrolysis, but also the packing procedure and the column hardware should be optimized towards operation in a temperature-programmed mode. This is a very important issue, because if the column hardware is made of components which will rapidly degrade due to temperature effects, then it is irrelevant if the stationary phase itself is stable at high eluent temperatures. Therefore, I would recommend running a few temperature gradients with a brand new column and comparing the column performance before and after the procedure.

A quick and easy test procedure to obtain information about the column stability under temperature-programmed conditions has been given in Chapter 5.1. This simple test allows the user to easily judge if a column is suitable for temperature-gradient operation or if problems will arise during method development. If the efficiency immediately decreases without a loss in retention, this may be related to an inappropriate packing procedure. The column hardware may also be composed of materials which are not sufficiently robust at high eluent temperatures. Often, leaks will be observed when the column is cooled down so that the fittings have to be re-tightened. In such instances, safe operation of the column is not possible and its usage cannot be recommended in high-temperature HPLC, even if the packing material is not prone to degradation. The XBridge column from Waters is a very good example of a column where both criteria are fulfilled: a stationary phase which is robust against hydrolysis and a column hardware which is suitable for high-temperature operation.

6.3 Mobile Phase Considerations

As has already been discussed, the mobile phase needs to be kept in the liquid state and phase transitions must be avoided. In Chapter 4, I have given some data on the vapour pressure of the mobile phase, so that the user is now able to decide how to adjust the back pressure and where to place the back-pressure restrictor. The complete data sets about the temperature dependence of vapour pressure, viscosity and static permittivity can be found in Appendices A, B and C at the end of this book. Please note that the back pressure can be adjusted either by using a back-pressure regulator or a restriction capillary. If you would like to use a restriction capillary, keep in mind that the viscosity of the mobile phase needs to be considered, because running a solvent and/or temperature gradient leads to significant changes in the viscosity of the mobile phase.

Generally, all the physicochemical parameters like vapour pressure, viscosity and static permittivity of the mobile phase change gradually as temperature is increased. This holds true as long as the boiling point line is avoided. There is only one exception, where unusual mixing behaviour is observed. This is the water–tetrahydrofuran system which exhibits a miscibility gap between 70 and 140 °C over a limited concentration range, provided that the pressure is below

247 bar.[6] If the miscibility gap is entered the two miscible solvents start to demix. If such a demixing occurs within the column, this poses a great problem for the reproducibility of retention times. Therefore, to avoid any problems with this solvent mixture, the miscibility gap has to be circumvented. This can be accomplished either by avoiding operation of the column between 70 and 140 °C, or by starting the solvent gradient at a high concentration of THF so that the water and tetrahydrofuran remain fully miscible even in this temperature region. Working with pure tetrahydrofuran, as is frequently done in gel-permeation chromatography, does not present a problem. You will find more information about this solvent system in Chapter 4.4, in which a strategy is presented of how to perform method development when working with a water–THF mixture.

A common question is "when does the mobile phase turn into a supercritical fluid?" As we have shown in a recent publication, for the six binary mobile phases given in Table 1.1 this can become possible as higher temperatures are approached.[7] For example, a mixture of water and ethanol at 250 °C turns into a supercritical fluid when the molar concentration of the organic modifier is above 80%. However, as long as you keep the temperature below 200 °C, there is no need to worry about this.

For precise back-pressure control, changes in the viscosity of the mobile phase need to be considered when solvent and/or temperature programming is applied. As was mentioned in Chapter 4, the overall system pressure depends not only on the pressure over the column, but also on the connecting capillaries. In order to correctly estimate the pressure which is generated by these connecting capillaries, you should disconnect the column from the system and then measure the pressure. You should then run the solvent gradient at the lowest and highest temperature you would like to apply and record the minimum and maximum pressures, respectively. Then you have a clear idea of the contribution of all the connecting capillaries to the system pressure. Please note that this pressure also depends on the flow rate, so you should adjust the flow rate to the highest value you would like to run the method. I would like to point out that it is very difficult to calculate the exact pressure drop over all capillaries. The reason is that – depending also on the type of oven you are using – only a small part of all capillaries will be heated. The rest of the tubing is just placed outside the oven. This means that the tubing which is placed outside the oven and goes from the injector to the column will not be heated. Regardless of whether you apply a temperature gradient or not, the temperature of the eluent in this capillary will remain constant. Then there are the capillaries which are directly located in the oven, which will be subjected to temperature change. Finally, all tubing which is placed outside the oven and goes from the column to the detector will also be influenced by temperature changes of the mobile phase. Please note that the hot mobile phase cools down when the tubing is not heated to a constant temperature and is exposed to ambient air. This means that there are also strong axial temperature gradients along this connecting tubing. As I mentioned, a complete mathematical treatment of the eluent temperature within those capillaries is very complex. Therefore, by performing an experiment as I have described here, you can be sure that all factors which directly influence the pressure drop in the tubing are included.

6.4 Influence of Temperature on Resolution

I guess that the aim of all chromatographic separation procedures is to optimize the resolution while minimizing the analysis time. Equation (6.1) can be regarded the master equation of resolution in liquid chromatography. I will discuss here the influence of temperature on resolution, as proposed by Knox and Thijssen:[i]

$$R = \left(\frac{k_1}{k_1 + 1}\right) \cdot (\alpha - 1) \cdot \frac{\sqrt{N}}{4} \tag{6.1}$$

This fundamental equation describes how resolution depends on three terms, which are: retention (k), selectivity (α) and efficiency (N). Each of these terms contributes to alter the resolution of the separation.[ii]

It is very interesting to note that all three terms can be influenced by temperature. This is not always realized even by experienced chromatographers. I will not give a full description on how method development is carried out in liquid chromatography, because this would be far outside the scope of this monograph. In this case, there are a number of excellent publications, as well as text books, which should be consulted. Dolan and Snyder have contributed a lot in the development of suitable software tools which can be used for method development and the interested reader is referred to the respective literature.[8–21] Recently, the authors published an updated version of a book on high-performance gradient elution.[22] The practitioner will find exhaustive information on how to proceed with structured method development. I can highly recommend this book for all who are interested in performing structured method development, which is not governed by "trial and error". In our laboratory, we also use computer optimization software, like DryLab® or ChromSwordAuto, which are indispensable tools for developing, optimizing and validating liquid chromatographic methods. It is clear that I cannot elaborate further on this, because a description of the complete theory, which is based on the Linear Solvent Strength model, would be far beyond the scope of this monograph. The intention of this chapter is to highlight how temperature can be incorporated into method development and how temperature affects

---

[i] Knox and Thijssen assumed equal peak width ($w_2 = w_1$) and consideration of the retention of the first peak of the critical pair. Other expressions were derived by Purnell or Said. Please note that you will find a lot of other "Master" equations, as they are frequently termed. In principle, they all contain the three terms which are also included in this equation.

[ii] It is very important to understand that the critical resolution is always the resolution of one critical peak pair. If a resolution 1.5 or 2 is obtained it means that for all other peak pairs in the mixture, a higher resolution results. Often, a resolution of 1.5 is said to be obligatory for a good separation, which means that all peaks are baseline resolved. However, sometimes this value is considered to be too low, especially if a small peak elutes after a large peak. In this case, a resolution of 2.0 or even higher may be required.

retention, selectivity and efficiency in liquid chromatography. I would therefore strongly advise the reader to consult the papers which I have listed above to become familiar with this concept, otherwise it may be difficult to follow the discussion in detail.

6.4.1 Influence of Temperature on Retention

I will now come to the question of how temperature affects retention. Usually in reversed-phase HPLC, retention will decrease if the temperature is increased. The influence of temperature on retention is given by the van't Hoff equation:

$$\ln k = -\frac{\Delta H}{R} \cdot \frac{1}{T} + \frac{\Delta S}{R} + \ln \beta \tag{6.2}$$

Here, ΔH is the enthalpy of transfer of the solute from the mobile into the stationary phase; ΔS is the entropy of transfer of the solute from the mobile into the stationary phase; R is the ideal gas constant; and β is the volume phase ratio of the stationary and mobile phase. However, this concept depends on a number of assumptions which may not always be met in reality. Often it is observed that the enthalpy of transfer is not a function of temperature. The same is assumed for the entropy of transfer, as well as the phase ratio of the column. At least theoretically, for most analytes a plot of the natural logarithm of the retention factor against the inverse absolute temperature (ln k *versus* $1/T$) yields a straight line, as depicted in Figure 6.1. The enthalpy of transfer can be calculated from the slope of this line, whereas the intercept will give the entropy of transfer of the solute from the mobile to the stationary phase.[iii] Please note that the vast majority of all compounds eluted in reversed-phase HPLC will show the behaviour depicted in Figure 6.1. This means that as the temperature is increased, the retention is decreased. If you carefully re-examine your chromatographic data, then you should observe the same trend. This is because in reversed-phase HPLC, the transfer of the analyte from the mobile phase to the stationary phase is usually exothermic.

Assuming that all analytes strictly obey the van't Hoff equation, a van't Hoff plot can be easily constructed where the retention of a compound is measured at two different temperatures. Both points can then be connected with a straight line and the retention can be predicted at any given temperature. Therefore, measuring the retention factor at two different temperatures is sufficient to describe the peak movement as the temperature is increased. It is the same concept which is incorporated into method development software which assists the practitioner to predict the retention of sample components as the temperature is varied.

However, care should be taken if a large temperature interval is considered, because in some cases there are deviations from linearity. As you will have

---

[iii] While it may be very interesting to calculate the thermodynamic properties of different solutes from retention time data, this information is practically useless in a routine laboratory where efficient method development is the primary goal.

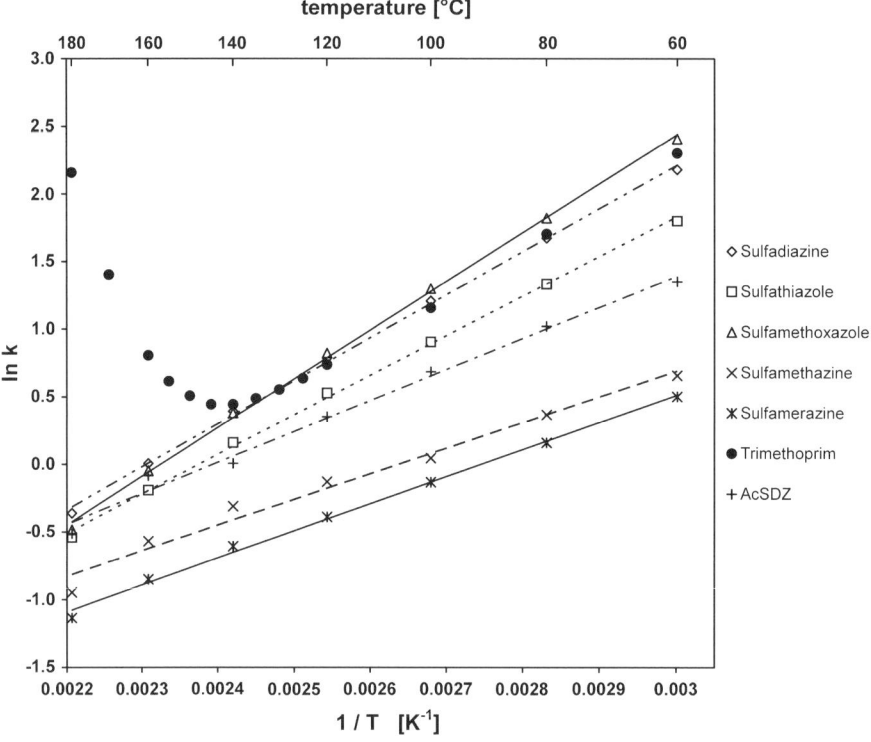

Figure 6.1 Monitoring of the retention of selected sulfonamides and trimethoprim on a carbon-clad zirconium dioxide stationary phase. Chromatographic conditions: column: ZirChrom-Carb (150 × 4.6 mm ID; 3 μm, 300 Å); mobile phase: deionised water–acetonitrile (containing 0.1% formic acid); flow rate: 1 ml min^{-1} (isocratic).[61] (Reproduced with kind permission from Elsevier).

noticed from Figure 6.1, trimethoprim does not follow this rule over the whole temperature range. This can be very helpful for adjusting the selectivity of the phase system, which I will outline in Chapter 6.4.2. To my best knowledge and experience, the concept of the temperature dependence of retention as described by the van't Hoff equation is most likely met if the interaction of the solute with the mobile and stationary phases is based on one single retention mechanism. In our laboratory, we often use polycyclic aromatic hydrocarbons in order to check the accuracy of retention time prediction. We have used this test mixture on several silica-based RP stationary phases and were able to predict retention times with a very small error when both the solvent gradient and the temperature are changed. The DryLab® software then creates a resolution map, where the critical resolution is calculated as the solvent gradient and the temperature are changed simultaneously. For these analytes, a perfect match is obtained, which means that the difference between experimental and predicted retention times is less than a second. This example

Table 6.1 Comparison between experimental and simulated retention times of a PAH mixture. Chromatographic details are given in the figure legend of Figure 2.1. Predictions have been made using the DryLab® 2000 Plus software.

Analyte	Experimental retention time [min]	Predicted retention time [min]	Difference [min]	Relative error [%]
1	0.47	0.48	0.02	3.2
2	0.56	0.57	0.01	2.4
3	0.72	0.73	0.02	2.4
4	0.78	0.79	0.01	1.8
5	0.91	0.90	−0.01	0.7
6	1.00	0.97	−0.03	3.2
7	1.06	1.01	−0.05	4.9
8	1.10	1.04	−0.05	4.9
9	1.19	1.12	−0.07	5.6
10	1.21	1.14	−0.07	5.7
11	1.29	1.22	−0.07	5.3
12	1.33	1.27	−0.06	4.8
13	1.38	1.33	−0.05	3.8
14	1.46	1.45	0.00	0.3
15	1.52	1.50	−0.02	1.3
16	1.58	1.59	0.01	0.5

highlights that the van't Hoff equation can be successfully used to predict retention times. I would like to refer the reader to Chapter 2, where a fast separation of a PAH mixture was shown (see Figure 2.1). This method was developed within one working day. The difference between experimental *versus* simulated retention time is given in Table 6.1. From these data it is obvious that only minimal differences are obtained.

It can be summarized that in reversed-phase HPLC, the effect of temperature on retention is determined by the retention enthalpy, which is usually negative as the retention factors of the sample components decrease with increasing temperature.

It is interesting to evaluate if the effect of this enthalpy change is different for small and large molecules. In 1993, Chen and Horváth published a paper in which they investigated the effect of temperature on nitrobenzene and lysozyme over a temperature range of 40 to 120 °C.[23] The results are depicted in Figure 6.2 and highlight that the slopes of the van't Hoff plots for nitrobenzene are much smaller than those for lysozyme. What is the practical meaning of this observation? The results demonstrate that the retention of large molecules can be influenced in a much stronger way than the retention of small molecules, if the same temperature range is considered. The conclusion is that temperature programming could be a real alternative to mobile phase gradients in the separation of macromolecules by reversed-phase HPLC.

In contrast to the examples I have given above, there are also deviations from the rule that a van't Hoff plot is always linear. It cannot be ignored that

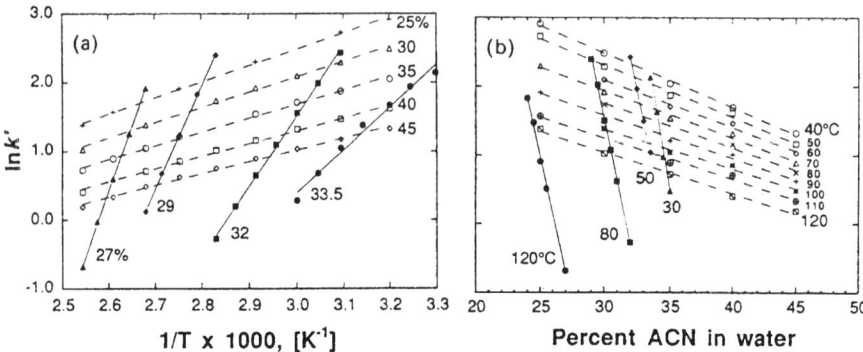

Figure 6.2 Plots of the logarithmic retention factor *versus* (a) the reciprocal absolute temperature or (b) % acetonitrile in water, in the reversed-phase chromatography of lysozyme (—) and nitrobenzene (- - -). The legends in (a) show the % acetonitrile in the aq. mobile phase, whereas those in (b) show the temperature for each set of data. Mobile phase containing 0.1% (v/v) trifluoroacetic acid was used in the chromatography of lysozyme. Column: (30 × 4.6 mm ID) packed with PLRP-S macroreticular crosslinked polystyrene. Flow rate: 1 ml min^{-1} (ref. 23). (Reproduced with kind permission from Wiley-VCH Verlag GmbH & Co. KGaA.)

temperature has a marked effect on pH and pK_a of analytes and the mobile phase. As was demonstrated by Castells *et al.*, the retention of ionic compounds can be influenced by changing the temperature leading to a shift in the pH of the mobile phase and hence in the degree of ionisation of the analytes.[24–26] This has also been demonstrated by McCalley *et al.*, who were able to show that for benzene a linear van't Hoff plot was obtained, while for the bases, amitriptyline, benzylamine, nortriptyline and quinine, retention increased with increasing temperature.[27] The authors assumed that this could be due to temperature-dependent pK_a shifts of the solutes. Very unusual retention behaviour was observed for protriptyline, whose retention decreased at temperatures between 30 and 45 °C and then increased at higher temperatures. So for some compounds it is not possible to get an accurate prediction of retention, due to secondary interactions of the analyte with the stationary and mobile phases. In such a case, I would suggest that a smaller temperature interval needs to be studied in order to minimize the error of the prediction. An alternative is to include more than two data points for the construction of the van't Hoff plot and use a polynomial function to model the temperature-dependent behaviour.

Apart from these aspects, a linear relationship between ln k and $1/T$ will only be obtained if the enthalpy and entropy of transfer are not a function of temperature. Furthermore, the phase ratio of the column has to be temperature independent. If a large temperature interval is considered, the volume occupied by the stationary phase might vary with the column average temperature and pressure.[28] This is because there is a thermal expansion of the column when the temperature is increased, as was already outlined in Chapter 5. If the column is operated at a high pressure, the compressibility of the bonded phase has to be taken into account.[iv,29] There are other reasons, however, why a deviation from linearity is observed.

Non-linear van't Hoff plots can also be explained by a phase transition of the stationary phase, which in most cases consists of ODS silica. The temperature at which a phase transition occurs is relatively low and lies at about 30 °C.[30] For the phase transition of the stationary phase, it is assumed that the ligands have a more rigid and hence structured order at temperatures below the phase transition. When the temperature is increased, the ligands can move more freely, which changes the interaction of the solute with the stationary phase. The overall conclusion which can be drawn is that curvilinear van't Hoff plots will likely result if there are two or more parameters which can be influenced by temperature in a different way. In this case, a severe deviation from the theoretically predicted model will be obtained. Principally, the van't Hoff equation is not limited to a certain temperature range. However, it is by no means advisable to define a certain temperature interval or a certain temperature range where linearity can be assumed without an experimental proof.

It should be obvious that temperature control is of utmost importance in order to obtain reproducible results, as I have already shown in Chapter 2 (see Figure 2.1). This evidence should convince the practitioner that it is also worthwhile to use temperature as an active variable in method development. I have to stress that, for the moment at least, every commercially available software tool which makes predictions of retention as a function of temperature, can only be used for isothermal operation of the column; the prediction of retention in the temperature-programmed mode is not currently possible. However, as will be shown in sections 6.5.2 and 6.5.3, temperature programming is an excellent tool for optimizing the selectivity of a separation. Additionally, most special hyphenation techniques, which will be introduced in Chapter 8, rely on temperature programming. Therefore, current activities in our laboratory are directed at developing and implementing new software tools for retention time prediction in temperature gradient mode. This feature will soon be available in the DryLab® software (distributed by the Molnár Institute for Applied Chromatography). The example which is depicted in Figure 6.3 highlights that a very accurate prediction of the retention time for a mixture of sulfonamides can be obtained if the elution is carried out under temperature-programmed conditions. The respective chromatogram is shown in Figure 6.4.

[iv] Chester and Coym presented a method to verify if a curvilinear van't Hoff plot was either caused by a change in the transfer enthalpy or a change in the phase ratio of the column. Instead of assuming that a curvilinear van't Hoff plot originates from a change in the transfer enthalpy, they constructed selectivity van't Hoff plots. This was accomplished by comparing the retention of two solutes differing structurally by one methylene unit, or by chromatographically analysing a homologous methylene series, and taking the slope of a plot of ln k vs. homolog number as the natural log of the methylene selectivity.

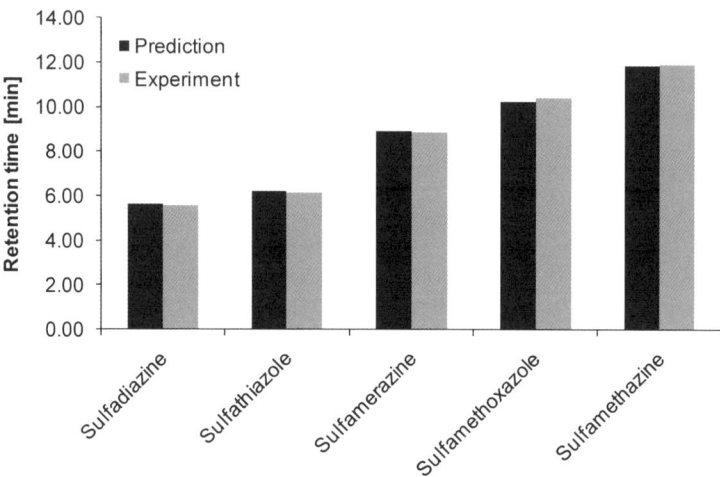

Figure 6.3 Comparison between experimental and simulated retention times of a sulfonamide mixture in temperature-gradient mode. (Chromatographic details are given in the legend of Figure 6.4.)

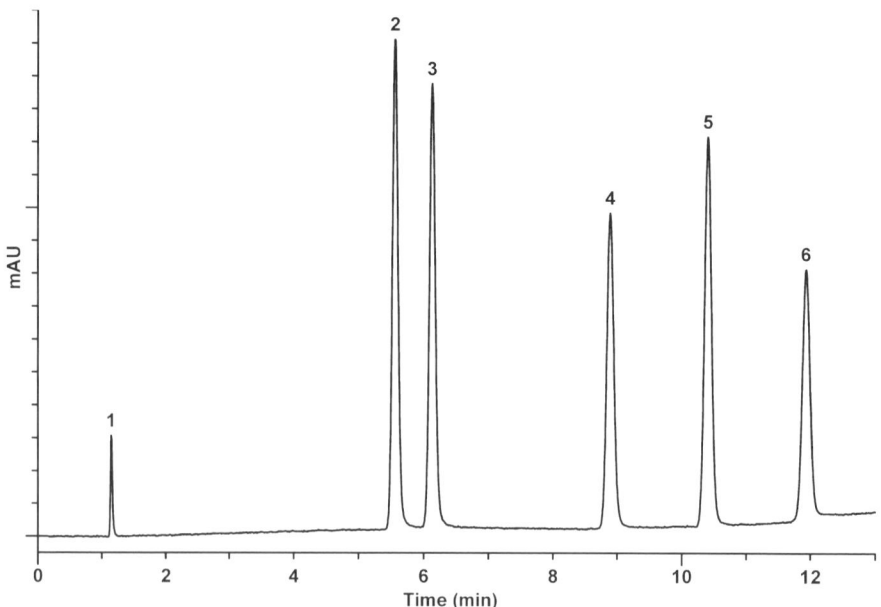

Figure 6.4 Temperature-gradient separation of five sulfonamides and uracil on a Waters XBridge BEH-C_{18} column (75 mm × 4.6 mm ID; 2.5 µm) with Shimadzu LC 10 and SIM HT-HPLC 200 column oven. Chromatographic conditions: mobile phase: water containing 0.1% formic acid; flow rate: 1.0 ml min^{-1}; injection vol.: 1.0 µl; detection: UV at 270 nm; temp. gradient: 60 to 115 °C in 4.4 min, and hold at 115 °C for 1.6 min, then 115 to 180 °C in 5.6 min, and hold at 180 °C for 0.9 min.

This tool will assist the user in finding the optimal temperature gradient for a given separation problem without arbitrarily varying the temperature. One reason why temperature programming is not used in a routine environment at the time of this writing is the missing capability of most advanced prediction software to model temperature gradients in liquid chromatography. With the incorporation of this tool into the DryLab® software, reliable predictions can be obtained. It should be noted that the same rules apply as for isothermal optimization, which means that non-linear van't Hoff plots due to secondary interactions may well be observed. Nevertheless, the commercial availability of this prediction capability should make temperature programming a valuable tool for method development and eliminate the problem of a mere "trial and error" approach.

I would like to close this section with an example I have already given in Chapter 5, where the concept of thermo-responsive stationary phases was discussed. As can be seen from Figures 5.9 and 5.10, the unusual behaviour where retention is increased with increasing temperature is due to the fact that the enthalpy change in these systems is endothermic. From Figure 5.10 it can also be derived that there are two regions for which a linear van't Hoff plot is assumed. This is a result of the phase transition of the stationary phase, which means that the polymer chains change their configuration at the lower critical solution temperature.[v] The example therefore serves to illustrate that method development should consider the phase transition of the stationary phase. Otherwise, a large error in the prediction of retention times of test solutes based on a van't Hoff plot, which is created by measuring the retention factor at 10 and 50 °C, would occur. In this case, a reliable prediction of retention times for intermediate temperatures would not be possible. It is also dangerous to extrapolate beyond the experimentally determined temperature range. This is immediately clear if you think about what happens when the retention model of the example in Figure 5.10, based on the temperature interval between 30 and 60 °C, is extrapolated to give a prediction window in the lower temperature range.

6.4.2 Influence of Temperature on Selectivity

The ability of a chromatographic system to discriminate between different analytes is called selectivity (α). Selectivity is determined as the ratio of the retention factors of two analytes, or the ratio of the reduced retention times and can be written as:

$$\alpha = \frac{k_2}{k_1} \tag{6.3}$$

---

[v] Please note that a phase transition here does not refer to the phenomenon of the mobile phase's transition from the liquid to the gaseous state, but instead refers to a conformational change of the stationary phase.

As was shown in the previous section, temperature has a significant influence on retention. Hence, temperature can also be used to control the selectivity of the separation, although in some text books it is stated that selectivity is not affected by temperature unless temperature modifies the nature of the analyte. This statement is untrue, and I will give some examples to highlight why temperature can be used to change the selectivity of the phase system. Usually, increasing the temperature leads to decreasing retention in reversed-phase HPLC, as I have shown in the previous section. Regardless of whether the analytes obey the van't Hoff equation or if a curvature is observed, the selectivity will change if the ratio of k_2 and k_1 is not constant with temperature. This means that even if all analytes obey the van't Hoff equation, a change in the separation factor may be obtained if the slopes of the van't Hoff plots for the selected analytes differ.

It is also wrong to claim that at higher temperatures, the selectivity will always decrease. Clearly, as I have shown in section 4.3, the static permittivity and thus the polarity of the mobile phase decreases, if the temperature is increased. This means that the elution strength increases if the temperature is increased, provided that the composition of the mobile phase is kept constant. In this case, the selectivity can also decrease up to the point where two peaks which are resolved at a low temperature will co-elute at a high temperature. Again, it strongly depends on whether a narrow or wide temperature interval is studied. A very good example to highlight these facts was presented by Edge *et al.*, who showed that for selected test probes, linear van't Hoff plots were observed between 40 and 180 °C using superheated water as the mobile phase, as shown in Figure 6.5.[31]

Although linearity was obtained for all compounds, a reversal of the elution order was noticed for caffeine and aminoantipyrine. Whilst separated at low temperature, the peaks coalesced at 113 °C but as the temperature was raised further they could again be separated. This example clearly demonstrated that for a given peak pair, selectivity decreases when the temperature is increased up to the point where there is a complete co-elution. However, if the temperature is increased further, a reversal of the elution order takes place and the selectivity then increases the higher the temperature is raised from the point of co-elution. *Therefore, temperature can be regarded as a very important parameter in influencing the selectivity of the phase system, despite the fact that selectivity is usually optimized by changing the mobile or stationary phases or by adjusting the solvent gradient.*

When a curvilinear van't Hoff plot for a compound is obtained, the change in selectivity may be even more pronounced. In our laboratory we were able to demonstrate this effect when a mixture containing selected sulfonamides and trimethoprim was analyzed on a carbon-clad zirconium dioxide stationary phase. While most of the analytes obeyed the van't Hoff equation, the retention of trimethoprim increased at temperatures above 140 °C. The results are illustrated in Figure 6.1. Although trimethoprim strongly overlapped with sulfadiazine in the temperature range between 60 and 140 °C, it moved away from all analytes when the temperature was increased further, and a dramatic increase

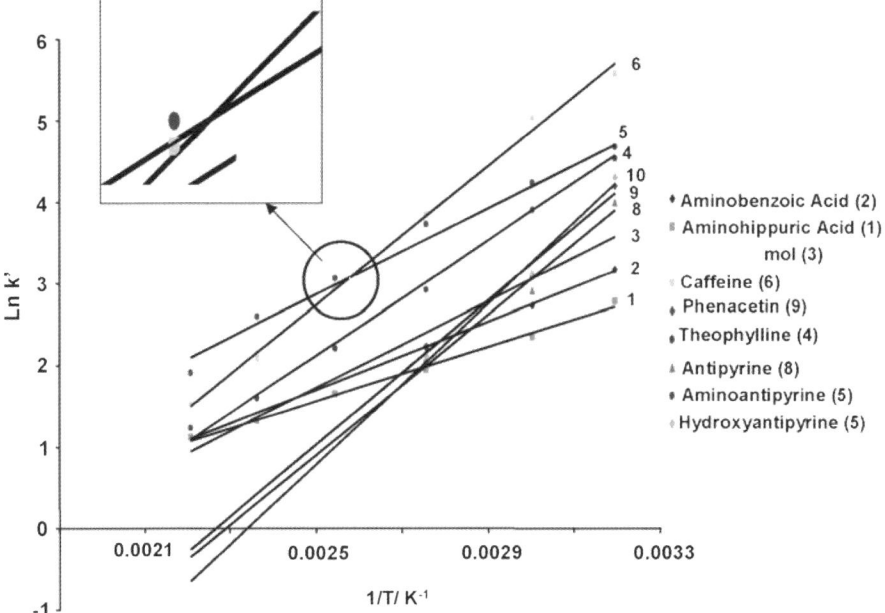

Figure 6.5 Van't Hoff plots for the test probes run on the Acquity column. The inset shows the point at which the elution order is reversed for aminoantipyrine and caffeine.[31] (Reproduced with kind permission from Elsevier.)

in the selectivity could be achieved. If the aim was to maximize resolution by means of selectivity, temperatures above 140 °C were clearly desirable. Hence, improving the selectivity can be a powerful tool for solving the separation problem of less complex samples.

The assumption that temperature plays only a minor role in controlling the selectivity can be attributed to studies that often examined only a very narrow temperature interval. The examples given in Figures 6.1 and 6.5 demonstrate that when a large temperature interval is considered, a reversal of the elution order between adjacent peaks is not unusual. In order to model the unusual retention behaviour of trimethoprim two linear regions can be assumed, as shown in Figure 5.10.

Since temperature exerts a strong influence on selectivity, the transformation of existing methods which are run at ambient temperature to a high-temperature separation is not always possible without additional experiments. Care should be taken that the effect of temperature on the quality of a separation should not only be reduced to kinetic aspects, which assumes that faster separations can be achieved at higher temperatures. Increasing the temperature in many cases also leads to a change in the selectivity of the separation. Therefore, the optimization of resolution by modulation of the selectivity of low complex samples is often superior to simply increasing the efficiency of the column by using smaller particles, or increasing the column length.

6.4.3 Influence of Temperature on Efficiency

Discussions on the influence of temperature on theoretical plate height are highly controversial. It is often stated that higher temperatures lead to higher efficiency. But is this statement really correct? The concept of the van Deemter equation was previously introduced in Chapter 1, where a very simple relationship was used. In order to investigate the question of the influence of temperature on efficiency in more depth, I have found a useful publication by Antia and Horváth from 1988, where this question has already been debated.[32] In that article, the authors derived an equation by combining the results of several authors, which looks a little bit more complicated than the simplified equation which was presented in Chapter 1:

$$H = A \cdot d_p + B \cdot \frac{D_m}{u} + \frac{C \cdot d_p^2}{D_m} \cdot u + \frac{D \cdot d_p^{5/3}}{D_m^{2/3}} \cdot u^{2/3} + \frac{2 \cdot k \cdot u}{(1+k)^2 \cdot k_d} \quad (6.4)$$

Here, A, B, C and D are dimensionless constants; D_m is the diffusion coefficient of the analyte in the mobile phase; k_d is the desorption rate constant; d_p is the particle diameter; and k is the retention factor of a solute (which was fixed at 3). In Equation (6.4), the first term arises from the poor distribution of the flow and is frequently called "eddy diffusion", depending on the quality of the column packing. It is generally assumed that this term does not depend heavily on temperature.[vi,33] The second term accounts for longitudinal molecular diffusion, which is dominant at very low flow velocities, because it depends on the residence time of the analyte in the column, but diminishes with increasing flow rate. The third term expresses the effect of resistance to intra-particulate mass transfer. The fourth term is indicative of resistance to mass transfer in the stagnant film of the mobile phase which surrounds the particles. The fifth term accounts for band spreading due to slow sorption kinetics. It is assumed that the parameters A, B, C and D do not change with temperature. Equation (6.4) can now be transformed to the reduced form by dividing the plate height (H) by the particle diameter (d_p) to obtain the reduced plate height (h), and by multiplying the linear velocity (u) with the particle diameter (d_p) and dividing this product by the diffusion coefficient (D_m) to obtain the reduced velocity (v):

$$h = A + \frac{B}{\nu} + C \cdot \nu + D \cdot \nu^{2/3} + \frac{3 \cdot D_m}{8 k_d \cdot d_p^2} \cdot \nu \quad (6.5)$$

All coefficients of the reduced velocity of the right hand side of this equation are independent of temperature, except the coefficient of the kinetic term. The reduced velocity itself, however, is dependent on temperature.

[vi] In this reference, the authors noted a slight decrease in A as the temperature is increased from 25 to 100 °C. This is attributed to the fact that there is an increase in the radial diffusion of solutes in the mobile phase, thereby opposing flow-induced dispersion.

Method Development using Temperature as an Active Variable 129

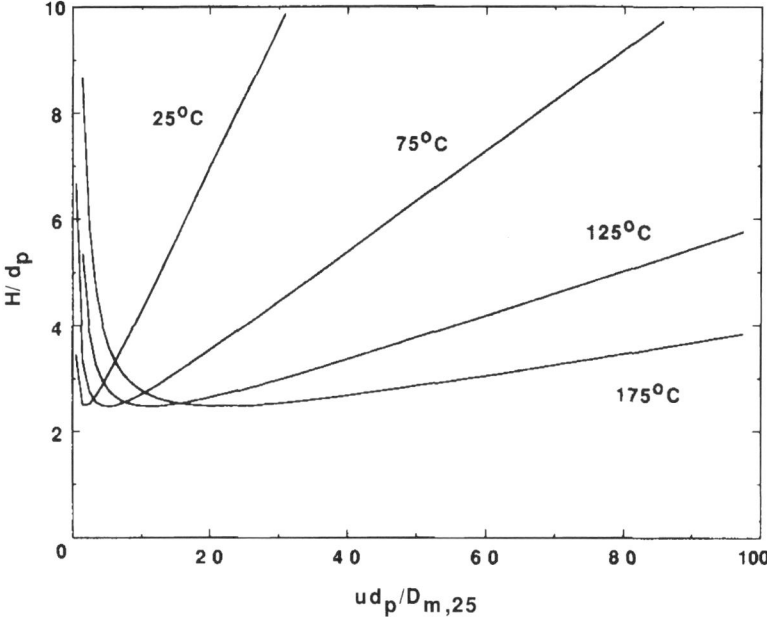

Figure 6.6 Plots of the reduced plate height *versus* the reduced velocity scaled to $D_{m,25}$ with the temperature as the parameter. Conditions: totally porous particles, rapid sorption kinetics, $D_{m,25} = 6 \times 10^{-7}$ and $d_p = 3\,\mu m$.[32] (Reproduced with kind permission from Elsevier.)

So, what does this now mean for the overall efficiency when temperature is increased? Chen and Horváth plotted the reduced plate height against the reduced velocity, assuming rapid sorption kinetics (see Figure 6.6).

As the temperature is increased, the optimum velocity also increases, although the minimum of the reduced plate height remains constant. However, at higher reduced velocities, the increase in the reduced plate height is much flatter than for low temperatures. While there is no absolute increase in the maximum efficiency, the loss in efficiency when the column is run at a much higher velocity than the optimum, which is usually observed at ambient temperature, is not very pronounced at high eluent temperatures. Please note that the plot which is depicted in Figure 6.6 is based on theoretical assumptions and exactly matches the experimental results which were depicted in Figure 1.1. In that figure, a plot of plate height *versus* linear velocity for propylparaben was shown, highlighting that at higher temperatures, the optimal linear velocity shifted to higher values while the absolute minimum of the *H*(*u*)-curve remained constant. Also, the flat shape of the van Deemter curve at higher eluent temperatures is clearly visible when operating the column at high flow rates. In this respect, there is a very good correlation between the experimental results and the theoretical prediction.

Now it can be understood why some authors derive different conclusions about the effect of temperature on efficiency. One of these studies has been conducted by Yang *et al.* who noted that the column efficiency is either improved or almost unchanged with increasing temperature in the temperature range between 60 and 120 °C, but is decreased when temperatures between 120 °C and 160 °C are applied.[34] This effect was the same for all columns which comprise a Zorbax RX-C8, a PRP-1 (polystyrene–divinylbenzene), a Hypersil ODS and a PBD-coated zirconium dioxide stationary phase. However, a different result was obtained in a related paper by the same authors.[35] In this study, the efficiency of test solutes on a PRP-1 column, a Chromatorex C-18 and a Zorbax RX-C18 column was recorded at different temperatures ranging from 60 to 140 °C. Here, plate numbers for all analytes always decreased with higher temperatures, except for catechol on the PRP-1 column when the temperature was increased from 100 to 140 °C. When the results are critically evaluated, it is immediately apparent that all the experiments were performed at constant flow rate. Therefore, in many cases at elevated temperatures, the flow rate was below the value for the maximum efficiency and an apparent loss in efficiency was observed as a result of molecular diffusion becoming a dominant factor. Hence, the results do not contradict the general conclusion which I have given above.

It can be summarized that the net benefit of operating HPLC columns at higher temperatures, therefore, is that the operator need not worry so much about the flow rate as long as it is higher than the optimum linear velocity. However, it needs to be stressed that there is no absolute increase in the efficiency, because it is not possible to lower the minimum of the van Deemter curve.[vii,36] It is common for such concepts to be incorrectly presented or misunderstood, such as when people speak of expecting an increase in efficiency by increasing temperature. However, these examples are valid when fast kinetics can be assumed. In systems with slow sorption kinetics, an increase in the temperature will accelerate the sorption, decreasing the value of the fifth term in Equation (6.5). Thus, the minimum plate height in systems with slow kinetics decreases with increasing temperature, and only approaches the value depicted in Figure 6.6 at higher temperatures.

Apart from this, there are other factors which are responsible for band broadening and a consequent loss in efficiency when the temperature is increased. The interested reader will find further information in a review written by Guiochon.[37] He gives a thorough discussion of the dependence of the HETP on column temperature as a result of radial temperature gradients formed inside the column.[viii] Please keep in mind that practical constraints of the heating system which were discussed in Chapter 3 might compromise efficiency, so that in effect a loss in efficiency might be observed at higher temperatures, as was already reported in a publication by Warren in 1988.[38]

[vii] A different result is obtained if secondary equilibria are concerned. De Villiers *et al.* presented a very interesting example where they studied the effect of temperature on the efficiency of a separation. They showed that for the separation of anthocyanins, an absolute increase in efficiency could be observed, which was due to secondary equilibria, where the compounds underwent a structural change on the column during the separation.

[viii] Detailed information about radial and axial temperature gradients has been given in Chapter 3.

Before I discuss method development strategies, I would like to address a few words in respect to the kinetic optimization of a separation, an approach which has gained broad acceptance. The vast majority of studies about kinetic aspects and efficiency have been conducted in the isocratic mode, which means a constant mobile phase composition.[39–49] In practice, however, we are mostly working in gradient mode. While for isocratic separations, efficiency is a good measure for describing the system performance, in gradient elution it is peak capacity. It should be noted that the optimal performance of a column in gradient elution is not necessarily linked with its operation at the van Deemter optimum, because this concept is only valid in isocratic mode. This suggestion was confirmed by Carr, who in 2006 raised this issue and stated that while these kinetic plots are undoubtedly useful for isocratic separations, their application in gradient elution is less straightforward.[50] He further elaborated that under solvent gradient conditions, peak capacity is the most common metric for assessing resolving power and used the term "conditional peak capacity".[ix] Here, a "discrepancy" is highlighted between isocratic and gradient elution. This refers to the gradient elution time, which depends on the complexity of the sample, as well as whether a simple or highly complex sample needs to be separated. What should be kept in mind is that in gradient elution, chromatographic bands can be compressed by the solvent gradient, which is not possible in isocratic elution. I think that the last word about the "optimal" method development strategy from a merely theoretical standpoint has not yet been delivered. This highly interesting, and sometimes controversial, discussion is continued in a paper which was published in 2009, which I would really like the interested reader to consult.[51] But now I will try to describe some practical approaches on how to select parameters to optimize resolution and speed in high-temperature HPLC. Please note that these strategies do not currently conform to a merely kinetic approach. The reason is that often technical difficulties prevail. I will come back to this problem in Chapter 8, where special hyphenation techniques are discussed. Moreover, a systematic method development has not been applied for the examples which are given in the following sections, because the necessity to implement new concepts for temperature optimization in liquid chromatography were derived in the last two years from the time of this writing.

6.5 Method Development

I will start with isothermal and isocratic separations and then gradually go over to more complex method development strategies, including a combination of solvent and temperature gradient programming.

6.5.1 Isothermal and Isocratic Separations

The first application I would like to discuss is the separation of steroids on a polybutadiene-coated zirconium dioxide stationary phase. Steroids are

---

[ix] Peak capacity is the measure of the number of peaks that can fit into an elution window t_1 to t_2 with a fixed resolution.

relatively non-polar and retention cannot be influenced by a pH adjustment. Therefore, the mobile phase does not need to be spiked with, for example, formic acid. The column we used had a length of 150 mm and an internal diameter of 4.6 mm. The particle diameter was 3 µm. The flow rate was adjusted to 1 ml min^{-1}, and detection was performed at 200 nm (since most of the target substances did not show a high UV activity at higher wavelengths). Figure 6.7 depicts the separation of a test mixture at isothermal and isocratic conditions. The temperature was adjusted to 25 °C and the mobile phase consisted of water–acetonitrile (75 : 25 v/v).

The resulting chromatogram does not look very exciting. All the compounds eluted with a strong tailing and the peak widths were excessively broad. Moreover, the analysis time was about 20 minutes. I have to emphasize that the column was purchased in 2001. This chromatogram reveals that the column shows a very low efficiency at ambient temperature, which is also well documented in the literature. Therefore, the column manufacturer recommended that these columns should only be used at elevated temperatures to maximize efficiency. It was assumed that the polymer layer was too thick so that the mass transfer was hindered. However, I have to repeat that the newer generation of zirconia columns does not exhibit such a poor performance at low temperatures. This is a clear sign that the column manufacturer has improved the

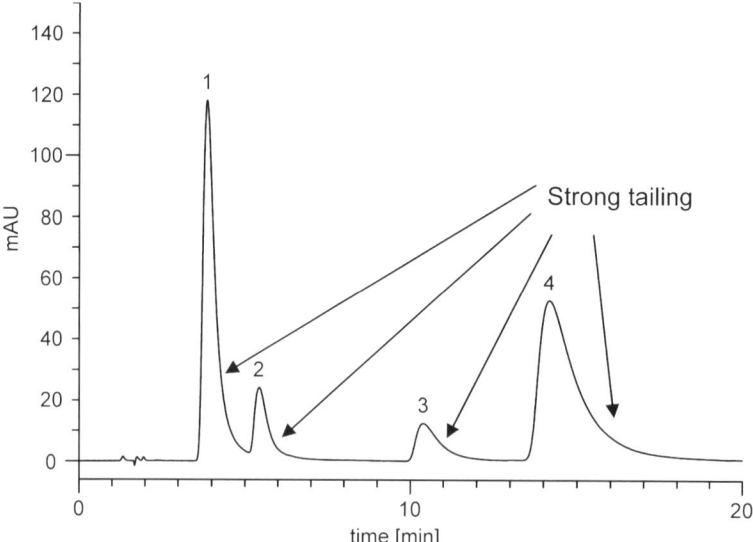

Figure 6.7 Isocratic and isothermal separation of a mixture containing four steroids on a polybutadiene-coated zirconium dioxide column (ZirChrom-PBD; 4.6 × 150 mm ID; 3 µm, 300 Å). Chromatographic conditions: temp.: 25 °C; flow rate: 1 ml min^{-1}; mobile phase: deionized water–acetonitrile (75 : 25 v/v); detection: UV at 200 nm. Peaks: 1, estriol; 2, androstadiendione; 3, dehydroepiandrosterone; and 4, estrone.[62] (Reproduced with kind permission from Elsevier.)

coating process, and the columns now show a comparable efficiency at low temperature to silica-based stationary phases. How can the separation now be optimized?

Increasing the temperature to 120 °C at a flow rate of 1 ml min^{-1} enables us to completely substitute the organic modifier with water, as is evident from Figure 6.8.

The separation now takes more than 30 minutes, but the resolution between adjacent peaks is very high so that the temperature can be further increased. As becomes obvious, by increasing the temperature to 140 and 185 °C, the analysis time can be reduced to 18 and 5.5 minutes, respectively. A comparison of the separation at 185 °C using only water as mobile phase, with the separation at 25 °C using a water–acetonitrile mixture (see Figure 6.7), shows that we were able to completely eliminate the organic content of the mobile phase while simultaneously reducing the analysis time by a factor of four. Since this is an isocratic and isothermal separation, there are no additional equilibration times once the temperature of the column is constant. Therefore, the system is ready for injection after 5.5 minutes and the analysis time corresponds to the cycle time of the complete method.

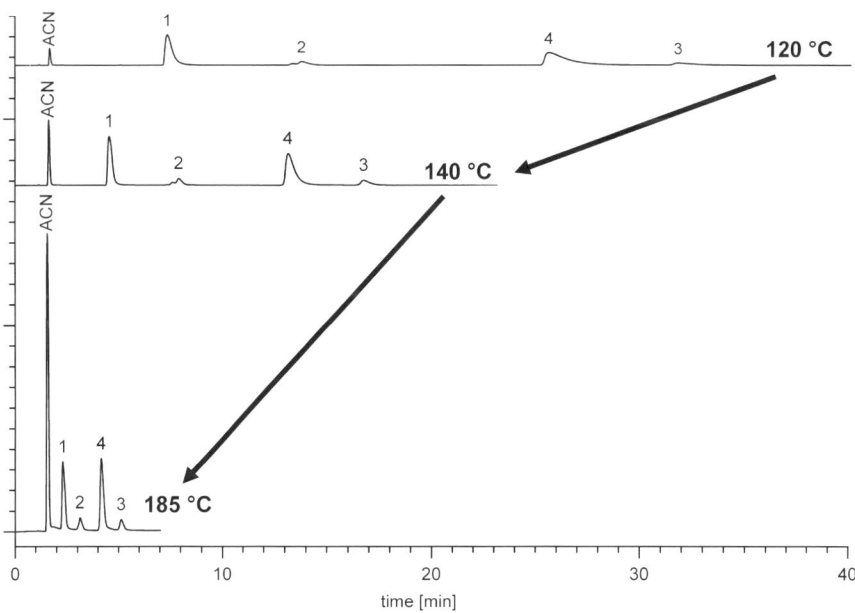

Figure 6.8 Isocratic and isothermal separation of a mixture containing four steroids on a polybutadiene-coated zirconium dioxide column (ZirChrom-PBD; 4.6 × 150 mm ID; 3 µm, 300 Å). Chromatographic conditions: temp.: 120, 140 and 185 °C; flow rate: 1 ml min^{-1}; mobile phase: deionized water; detection: UV at 200 nm. Peaks: 1, estriol; 2, androstadiendione; 3, dehydroepiandrosterone; and 4, estrone.

We can speed up analysis even further. At 185 °C, a significant increase in flow rate can be achieved due to the lower viscosity. We were able to increase the flow rate to 5 ml min^{-1} (without exceeding the recommended pressure limit of this column at 300 bars), which led to an additional five-fold reduction in analysis time. The resulting chromatogram is depicted in Figure 6.9.

A close inspection of the peaks reveals that the temperature of the mobile phase matched the temperature of the stationary phase, because all the peaks eluted symmetrically and no peak splitting was observed. What is even more important is that the increase in flow rate led to a proportional decrease in the retention time for all analytes. On doubling the flow rate, retention was halved. In other words, if there had been a so-called "thermal mismatch" between the eluent entering the column and the stationary phase, radial and axial temperature gradients would have built up in the column as was demonstrated in Figure 3.1 (see Chapter 3). However, the absence of peak splitting and the respective decrease in the analytes' retention times, unambiguously confirm that the mobile phase preheating is adequate and also effective at high flow rates up to 5 ml min^{-1}. It must be noted that a flow rate of 5 ml min^{-1} is very high, taking into account that UV detection is replaced more and more by mass spectrometric detection. Usually, the ion sources of these instruments are not designed for high flow rates. Especially electrospray ionization (ESI) requires

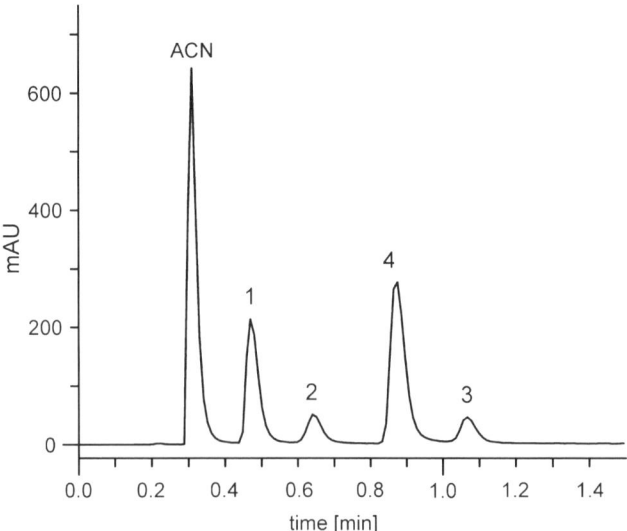

Figure 6.9 Isocratic and isothermal separation of a mixture containing four steroids on a polybutadiene-coated zirconium dioxide column (ZirChrom-PBD; 4.6 × 150 mm ID; 3 µm, 300 Å). Chromatographic conditions: temp.: 185 °C; flow rate: 5 ml min^{-1}; mobile phase: deionized water; detection: UV at 200 nm. Peaks: 1, estriol; 2, androstadiendione; 3, dehydroepiandrosterone; and 4, estrone.[62] (Reproduced with kind permission from Elsevier.)

Method Development using Temperature as an Active Variable 135

low flow rates in the range of 300 to 500 µl min^{-1} to achieve optimal ionization efficiency, while atmospheric pressure chemical ionization (APCI) might be incompatible with flow rates over 2 ml min^{-1}. Thus, flow rates up to 5 ml min^{-1} are too high for MS detection, unless a flow splitter is used so that only a fraction of the mobile phase is introduced into the mass spectrometer. Even for UV detection, this is a very high flow rate, leading to a high consumption of solvents. Since in this example only water was used as the mobile phase, this did not matter. The main intention of this example, therefore, was to demonstrate that such high flow rates can be used without a significant loss in efficiency.

In Chapter 8, I will continue with this example and elucidate the concept of high-temperature HPLC hyphenated to isotope ratio mass spectrometry. In this context it will then become clear as to why the use of a polymer-coated metal oxide stationary phase is a very good choice for hyphenation techniques which rely on the use of a pure water mobile phase.

6.5.2 Temperature Gradient and Isocratic Separation

In this section, I will explain the concept of temperature programming and show that a solvent gradient can be replaced by a temperature gradient. The first example covers a mixture which contains compounds of differing polarity. In this mixture, cytarabine is the most polar compound which is often excluded from the pores of an RP stationary phase and therefore elutes before the void time. In contrast to this, etoposide is relatively non-polar, eluting as the last compound in this mixture. The method was developed at 25 °C on a polystyrene-coated zirconium dioxide column. An isocratic mobile phase of water and acetonitrile was used, which was spiked with 0.1% formic acid. The flow rate was adjusted to 1 ml min^{-1} and a 150 × 4.6 mm column packed with 3 µm particles was used. The resulting chromatogram depicted in Figure 6.10 reveals that although the polar compounds are nicely separated, etoposide elutes with a pronounced tailing around 21 minutes. This means that a lot of time is wasted, because the resolution between chloramphenicol and etoposide is too high.

In order to optimize the separation without applying a solvent gradient, the only option is to increase the temperature so that the retention of etoposide can be decreased. Figure 6.11 demonstrates what happens if the temperature is increased to 80, 100 and 130 °C, while the composition of the mobile phase remains constant.

It can be seen that the resolution of the peaks eluting at the beginning of the chromatogram continually worsens, while at 130 °C an adequate resolution is obtained between chloramphenicol and etoposide. Also, the higher temperature leads to an improvement in the signal-to-noise ratio, because the excessive tailing which was noticed for etoposide at low temperatures has now been reduced. However, we are far away from a good separation, because a quantification using UV detection at 254 nm cannot be made for the early eluting compounds due to co-elution.

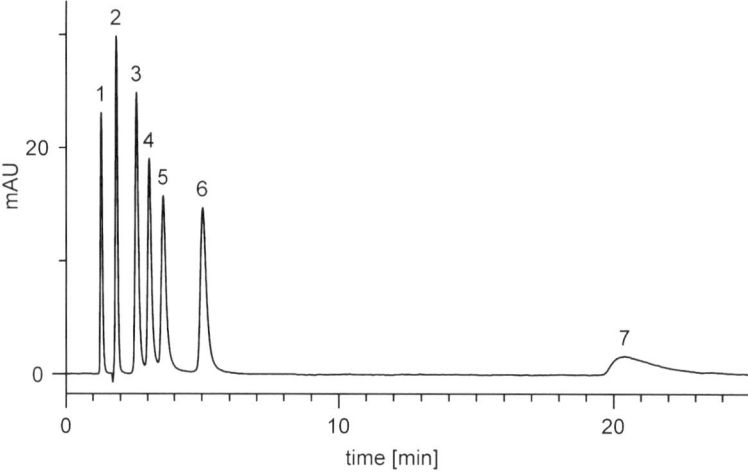

Figure 6.10 Isocratic and isothermal separation of a mixture containing cytostatics and antibiotic drugs on a polystyrene-coated zirconium dioxide column (ZirChrom-PS; 4.6 × 150 mm ID; 3 µm, 300 Å). Chromatographic conditions: temp.: 25 °C; flow rate: 1 ml min^{-1}; mobile phase: deionized water–acetonitrile (90 : 10 v/v) containing 0.1% formic acid; detection: UV at 254 nm. Peaks: 1, cytarabine; 2, 5-fluorouracil; 3, sulfadiazine; 4, sulfathiazole; 5, sulfamethoxypyridazine; 6, chloramphenicol; and 7, etoposide.[62] (Reproduced with kind permission from Elsevier.)

In order to improve the resolution while minimizing the overall analysis time, temperature programming was applied. Since temperature has a pronounced impact on the polarity of the mobile phase, as was discussed in Chapter 4.3, it is possible to use pure water spiked with 0.1% formic acid. The start temperature of the gradient was set at 40 °C, while the upper temperature was set at 130 °C. Within four minutes, the temperature was raised linearly from the lower to the upper temperature. As can be seen from Figure 6.12, the polar compounds are well resolved and etoposide now elutes at approximately 5.5 minutes. When compared with the isothermal separation at 25 °C, the analysis time was reduced by a factor of four. A much faster elution was achieved since the flow rate was increased to 2 ml min^{-1}. Again, a higher flow rate is beneficial when working at higher temperatures due to a much flatter increase in the van Deemter curve in the C-term dominated region.

In this context we should differentiate between the analysis time, which is measured from the point of injection to the point where the last compound elutes from the column, and the cycle time, which also takes into account the time necessary for re-equilibration. In solvent gradient elution, time is always needed for re-equilibrating the column when the mobile phase is changed after the solvent gradient. This is the same for temperature programming. Although we are working under isocratic conditions, the time which is needed to thermally equilibrate the column must be added to the overall analysis time. How does it work? As I mentioned in Chapter 3.2, a block-heating oven is usually

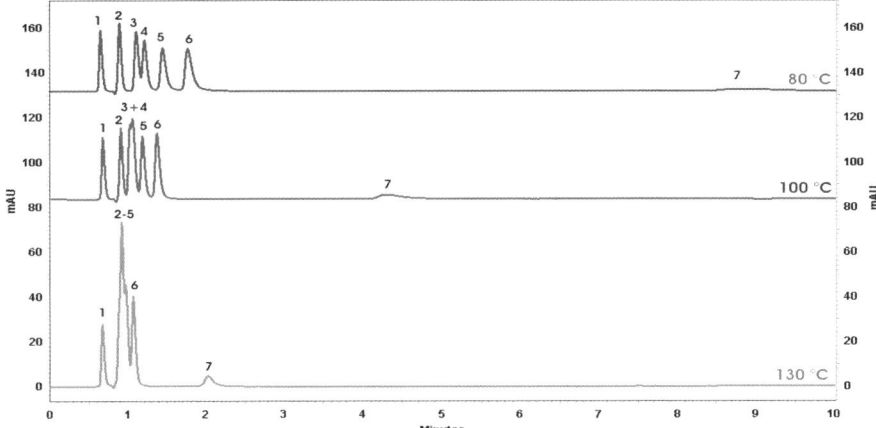

Figure 6.11 Isocratic and isothermal separation of a mixture containing cytostatics and antibiotic drugs on a polystyrene-coated zirconium dioxide column (ZirChrom-PS; 4.6 × 150 mm ID; 3 µm, 300 Å). Chromatographic conditions: temp.: 80, 100 and 130 °C; flow rate: 2 ml min^{-1}; mobile phase: deionized water–acetonitrile (90 : 10 v/v) containing 0.1% formic acid; detection: UV at 254 nm. Peaks: 1, cytarabine; 2, 5-fluorouracil; 3, sulfadiazine; 4, sulfathiazole; 5, sulfamethoxypyridazine; 6, chloramphenicol; and 7, etoposide.

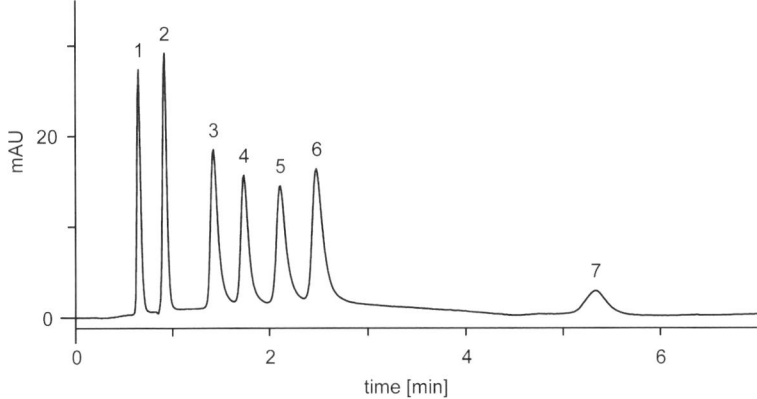

Figure 6.12 Isocratic and temperature-programmed separation of a mixture containing cytostatics and antibiotic drugs on a polystyrene-coated zirconium dioxide column (ZirChrom-PS; 4.6 × 150 mm ID; 3 µm, 300 Å). Chromatographic conditions: temp. gradient: 40 to 130 °C in 4 min; flow rate: 2 ml min^{-1}; mobile phase: 100 % deionized water (containing 0.1% formic acid); detection: UV at 254 nm. Peaks: 1, cytarabine; 2, 5-fluorouracil; 3, sulfadiazine; 4, sulfathiazole; 5, sulfamethoxypyridazine; 6, chloramphenicol; and 7, etoposide.[62] (Reproduced with kind permission from Elsevier.)

considered to be inappropriate for temperature programming. The reason is that a large thermal mass has to be cooled down before the system is ready for injection. However, the operating principle of our system allows for a rapid cooling down of the mobile and stationary phases after the temperature gradient, by using an internal cooling cycle which is in direct thermal contact with the heating elements. The system was ready for injection after a total cycle time of 10 minutes. In order to prove that constant conditions are achieved, some important parameters, including retention time (t), peak area (A) and the quotient of A/t in 10 consecutive runs, are listed in Table 6.2 for all analytes.

From this data the mean, the standard deviation and the relative standard deviation (RSD) were calculated. The results underline that the repeatability of the method is very good. However, it has to be acknowledged that due to the thermal re-equilibration, the cycle time is only reduced by a factor of two and not by a factor of four when compared to the initial isothermal chromatogram in Figure 6.10. Nevertheless, a two-fold reduction in analysis time and a concomitant reduction in solvent waste lead to a considerable turnover of the laboratory's efficiency. Because of the influence of temperature on the polarity of the eluent, the separation can also now be carried out without using acetonitrile as an organic co-solvent. Although discussions concerning the reduction in toxic and costly organic solvents are often controversial, the worldwide shortage of acetonitrile in 2009 has led to new thinking with regard to low-cost, readily available mobile phases. Clearly, the final method (see Figure 6.12) fulfils these requirements, as water is environmentally friendly and can be purchased at a low cost for a high purity. Moreover, the analysis time could be reduced so that a higher throughput can be achieved. Hence, this example highlights that temperature programming can indeed be used to achieve an increase in the laboratory's efficiency, despite the thermal re-equilibration of the column which was not necessary when working isothermally.

One aspect, however, needs to be discussed in more depth. When the chromatogram in Figure 6.12 is carefully examined, it is obvious that etoposide elutes as a relatively broad peak, because a band compression is not possible. In solvent gradient mode, a band compression can be achieved which reduces peak broadening. Why is this? When a solvent gradient is considered, a band migrates through the column in a mobile phase of continuously increasing strength. As a result, the tail of the band always moves in a mobile phase that is slightly stronger than the mobile phase at the front of the band. The tail therefore tends to move faster than the front. The resulting narrowing of the band is referred to as "gradient compression", which serves to partly counteract the normal broadening of the band during migration through the column. The interested reader will find more information in the respective literature.[52–54] A different result is obtained if a temperature gradient is applied. In this case, the temperature immediately affects the whole band. This also means that a compression is not observed when the column is operated isocratically under temperature-programmed conditions. It could be argued that it would be advantageous to simultaneously increase the flow rate when temperature programming is performed. In principle, there is nothing which speaks

Table 6.2 Retention time t, peak area A and the quotient of A/t from 10 consecutive chromatographic runs under temperature-programmed conditions as specified in Figure 6.12.

Compound		Run 1	Run 2	Run 3	Run 4	Run 5	Run 6	Run 7	Run 8	Run 9	Run 10	Mean	SD	RSD/%
Cytarabine	RT/min	0.640	0.651	0.651	0.651	0.640	0.651	0.651	0.640	0.651	0.640	0.647	0.0057	0.88
	Peak Area	73661	73661	74086	73513	72937	73069	72922	73289	72806	72808	73275	439	0.60
	Area/RT	115095	113151	113803	112923	113964	112241	112015	114514	111837	113763	113331	1089	0.96
5-Fluorouracil	RT/min	0.907	0.917	0.917	0.917	0.907	0.917	0.917	0.907	0.917	0.907	0.913	0.0052	0.57
	Peak Area	90324	88326	89805	88530	88001	87709	87875	88141	87779	88351	88484	882	1.00
	Area/RT	99585	96321	97933	96543	97024	95648	95829	97179	95724	97410	96920	1211	1.25
Sulfadiazine	RT/min	1.419	1.429	1.419	1.419	1.408	1.419	1.419	1.408	1.419	1.408	1.417	0.0068	0.48
	Peak Area	94191	93319	93660	93209	92601	92592	92996	93025	92380	92772	93075	548	0.59
	Area/RT	66378	65304	66004	65686	65768	65252	65536	66069	65102	65889	65699	404	0.62
Sulfathiazole	RT/min	1.739	1.739	1.728	1.728	1.717	1.728	1.728	1.717	1.728	1.717	1.727	0.0081	0.47
	Peak Area	76080	76732	76848	76756	75929	75462	76224	76723	76351	76264	76337	442	0.58
	Area/RT	43749	44124	44472	44419	44222	43670	44111	44684	44185	44417	44205	317	0.72
Sulfamethoxy-pyridazine	RT/min	2.112	2.112	2.101	2.101	2.091	2.091	2.091	2.080	2.091	2.080	2.095	0.0114	0.54
	Peak Area	85944	87348	86603	86580	86271	85165	85801	85731	86476	86337	86226	599	0.70
	Area/RT	40693	41358	41220	41209	41258	40729	41033	41217	41356	41508	41158	266	0.65
Chloramphenicol	RT/min	2.475	2.475	2.464	2.464	2.453	2.453	2.453	2.443	2.453	2.443	2.458	0.0115	0.47
	Peak Area	114960	114370	115427	114055	112978	113651	113380	114574	114566	112833	114079	856	0.75
	Area/RT	46448	46210	46845	46289	46057	46331	46221	46899	46704	46186	46419	296	0.64
Etoposide	RT/min	5.344	5.333	5.312	5.301	5.280	5.280	5.280	5.259	5.269	5.248	5.291	0.0313	0.59
	Peak Area	35376	38530	36344	35239	36827	37474	37588	36175	37365	37925	36884	1088	2.95
	Area/RT	6620	7225	6842	6648	6975	7097	7119	6879	7091	7227	6972	220	3.15

against this advice. A problem might only arise if the flow rate becomes too high for certain detectors. What might be a good alternative is to use capillary columns, because the flow rate is compatible to mass spectrometric detection and the columns can also be operated at a high linear velocity without the need to use flow splitters. Another advantage for using capillary columns is that the problem of radial temperature gradients is much less pronounced, although it was demonstrated in Chapter 3 that an efficient heat transfer can be achieved for normal-bore columns when a block-heating device is used. *It can be summarized that an effective band compression is not achieved in the isocratic mode when a temperature gradient is applied.* I will show some examples in Chapter 8 where the use of organic solvents is strictly forbidden, so that temperature programming remains the only option for changing the selectivity and the elution strength of the mobile phase.

6.5.3 Simultaneous Temperature and Solvent Gradient Separation

A concept I would like to revisit in this section is based on the simultaneous use of solvent and temperature programming. A solvent gradient which is applied simultaneously to a temperature gradient may be beneficial in terms of improving efficiency, by means of a band compression. This means that for applications where the use of an organic co-solvent is not "prohibited", solvent-gradient programming should always be applied in combination with temperature programming. As was shown in Chapter 6.4.2, selectivity can be a powerful guide for optimizing the chromatographic resolution. In many cases, the separation of critical peak pairs is not possible even if the number of theoretical plates is increased by using smaller particle packed columns. Only an increase in selectivity will lead to an improvement in resolution. It is useful to screen the literature for publications describing the concept of simultaneous solvent and temperature programming. Again, the great man of chromatography, Csaba Horváth, laid the groundwork in his pioneering work. In 1997, Chen and Horváth demonstrated that temperature programming can be used in combination with solvent programming.[55] They showed that by applying a simultaneous temperature programme starting from 30 °C and using a heating rate of 10 °C min^{-1}, the best compromise between resolution and analysis time was achieved for a given separation problem when compared with solvent elution at 30 and 80 °C.

Very recently, we adapted this approach for the separation of some sulfonamides and trimethoprim. The goal was to establish a fast method where all target analytes were at least baseline resolved. The equipment which was available for method development consisted of an Agilent HPLC system with a maximum operating pressure of 400 bars. In order to achieve a fast elution, the temperature had to be increased to decrease the viscosity of the mobile phase. First of all we performed isothermal measurements at 70 and 90 °C. However, as can be seen from Figuress 6.13b and c, neither the separation at 70 °C nor the

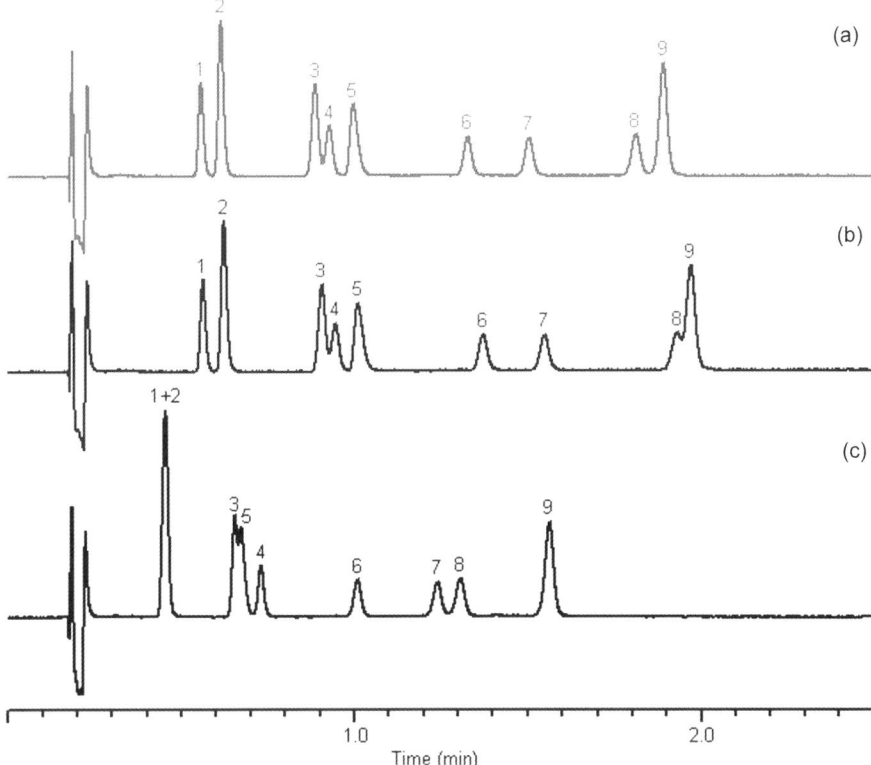

Figure 6.13 Chromatograms showing the separation of eight sulfonamides and trimethoprim with a solvent gradient (a) and a temp. gradient from 70 to 90 °C in 2 min, (b) at 70 °C (isothermal) and (c) at 90 °C (isothermal). Injection vol.: 1 µl of a 500 µg ml^{-1} sample. Peaks: 1, sulfadiazine; 2, sulfathiazole; 3, N^4-acetylsulfadiazine; 4, sulfamerazine; 5, trimethoprim; 6, N^4-acetylsulfamerazine; 7, sulfamethazine; 8, sulfamethoxazole; and 9, N^4-acetylsulfamethazine. Chromatographic conditions: flow rate: 1.4 ml min^{-1}; solvent A: deionized water (containing 0.1% formic acid); solvent B: acetonitrile (containing 0.1% formic acid); solvent gradient: 7 to 15% solvent B in 2.5 min.[63] (Reproduced with kind permission from Wiley-VCH Verlag GmbH & Co. KGaA.)

separation at 90 °C were successful. At the lower temperature, the last peak pair was not fully resolved, while at the higher temperature, co-elution of the first peaks could be observed. Hence, a temperature gradient was applied linearly increasing from 70 to 90 °C, since at the lower temperature a better resolution is achieved for the first peaks, while the higher temperature is beneficial for the last eluting compounds. The resulting chromatogram is depicted in Figure 6.13a.

Please note that it was possible to speed up the separation even further by adjusting the solvent and temperature gradient. Hence, both gradients were optimized by applying two-step gradients. The question might be asked whether

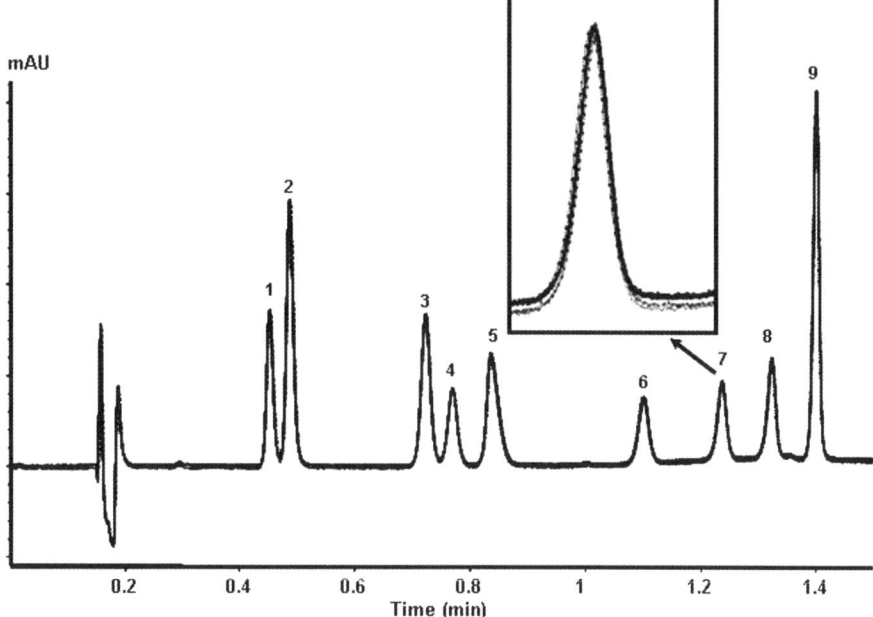

Figure 6.14 Overlay of nine chromatograms showing the separation of eight sulfonamides and trimethoprim with simultaneous solvent and temperature gradients. Peaks: 1, sulfadiazine; 2, sulfathiazole; 3, N^4-acetylsulfadiazine; 4, sulfamerazine; 5, trimethoprim; 6, N^4-acetylsulfamerazine; 7, sulfamethazine; 8, sulfamethoxazole; and 9, N^4-acetylsulfamethazine. Chromatographic conditions: flow rate: 1.7 ml min^{-1}; solvent A: deionized water (containing 0.1% formic acid); solvent B: acetonitrile (containing 0.1% formic acid); solvent gradient: 6 to 8% solvent B in 0.4 min, then 8 to 40% solvent B in 1.1 min; temp. gradient: 80 to 90 °C in 1.1 min, then 90 to 105 °C in 0.4 min.[63] (Reproduced with kind permission from Wiley-VCH Verlag GmbH & Co. KGaA.)

the retention times can be precisely reproduced in consecutive runs, or whether it was a mere accident that the optimal resolution was achieved with a concomitant temperature gradient. Therefore, Figure 6.14 shows an overlay of nine consecutive runs, highlighting that a shift in retention times cannot be observed.

This example highlights that the simultaneous use of solvent and temperature programming can significantly enhance the selectivity of the separation. Even small changes in temperature can significantly influence peak movement. Of course, this is an advanced technique and temperature programming should only be performed if the heating system allows for a fast temperature change. Indeed, there are other heating systems which, in principle, could also be used for temperature programming, but you have to keep in mind that the cooling of the mobile phase can take as long as 30 minutes. In this case, speaking of fast chromatography would be ridiculous and it makes no practical sense to apply temperature programming.

Method Development using Temperature as an Active Variable 143

I would like to close this section with a few words addressing systematic method development in high-temperature HPLC using temperature programming. As was already outlined, whenever possible, computer optimization software should be used to model the retention of the sample compounds. At the time of writing, work is currently being carried out in our laboratory to implement temperature-gradient optimization for method development in high-temperature RPLC, as I have already described in section 6.4.1. The first results are quite encouraging, so that in the near future a commercially available solution should be available.[x] It will then be possible to carry out a systematic method development without the problem of "trial and error" optimization. Clearly, the chromatograms I have shown in this chapter have been developed by a "trial and error" approach, considering the theoretical assumptions outlined in sections 6.4.1 to 6.4.3. However, it is often not possible for a human operator to find the best chromatographic conditions by simply changing the gradient slope arbitrarily. This means that it can be a very time-consuming and tedious procedure to arrive at the final method, which might not even be the optimum method. With the help of computer optimization software, a resolution map is generated which enables the best parameters for optimizing resolution and minimizing analysis time to be found. Although temperature programming is regarded too specialized to be of any practical value in most routine laboratories, the example depicted in Figure 6.13 unambiguously reveals that even small changes in temperature can greatly affect selectivity. In this respect, the availability of advanced software packages will be of great help to the practitioner, who may otherwise waste precious time without necessarily finding the optimum separation conditions.

6.5.4 Detector Optimization

While the influence of temperature on chromatographic parameters has been shown to be very important in method development, the influence of temperature on the signal-to-noise ratio also needs to be considered. In Chapter 3.3, I emphasized that a post-column cooling unit might be essential for detectors where temperature has a negative impact on the signal. This is true for the fluorescence detector, where the signal can be quenched if the inlet temperature is increased. In contrast to this, a gain in signal intensity might be achieved for detectors where the liquid mobile phase is converted to a gas during the detection process, which holds true for the evaporative light-scattering detector (ELSD), the charged aerosol detector (CAD) and the mass spectrometer. Based on the same experiment that I described in Chapter 3.3 (see Figure 3.12), the results given here are for ELS-detection (see Figure 6.15).

It can be seen that a tremendous gain in peak area results when the temperature of the mobile phase is increased. Why is that? First of all, the mobile phase has to be converted to a gas after it is introduced into the ELSD. Therefore, it seems to be advantageous if temperature is very close to the

---
[x/]This work is a close collaboration with the Molnár Institute for Applied Chromatography.

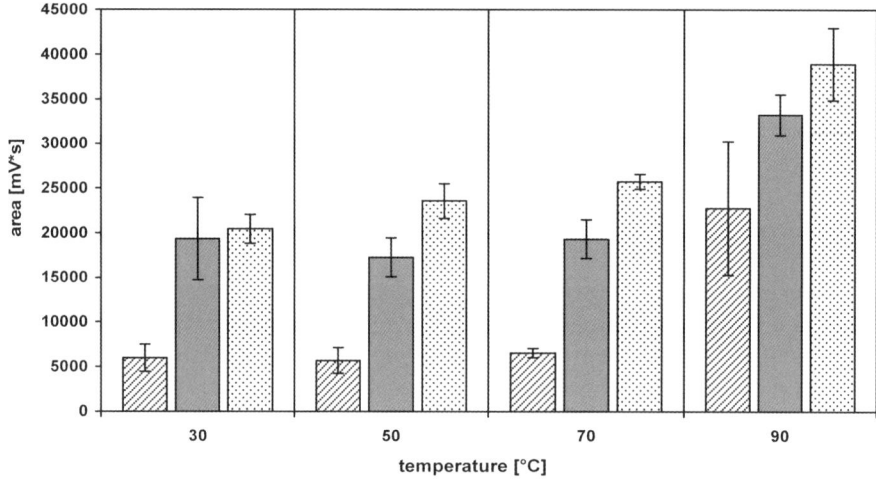

Figure 6.15 Peak area of three PAH at different eluent temperatures. Chromatographic conditions: mobile phase: deionized water–acetonitrile containing 0.1% formic acid (50 : 50 v/v); flow rate: 1 ml min^{-1}; detection: ELSD (nebuliser temp.: 40 °C; evaporation temp.: 90 °C).

transition temperature of the mobile phase, supporting the phase transition from liquid to gaseous. During this process, a gas flow of air or nitrogen is mixed with the effluent to produce an aerosol of minute droplets. Hence, with the help of a heated transfer line, a higher evaporation efficiency can be achieved.

A further characteristic of these types of detectors is that the signal also depends on the composition of the mobile phase. Running a solvent gradient from 5 to 100% B means that not only the composition of the mobile phase changes, but other important parameters, like viscosity or surface tension, are also affected which can alter the nebulisation stage. This in turn can lead to an enhancement of the signal during the solvent gradient when the same compound is injected into the gradient. One solution to this problem could be mobile phase gradient compensation which has been described by Gorecki and de Villiers both ELSD and CAD.[56,57] Another approach, however, could be to use temperature programming at isocratic conditions. In this case, the composition of the mobile phase is kept constant. An experimental set-up of a heated transfer line between the column and the detector is described in Chapter 4 (see the description of sputtering effect in Chapter 4.1.2) which should help to keep all the parameters constant. With this technique, a constant inlet temperature of the mobile phase can be achieved and a phase transition of the effluent will be avoided.

The same effect has been shown by Heinisch and co-workers for ESI-MS.[58] They investigated the impact that temperature has on the signal-to-noise (S/N) ratio when a high eluent inlet temperature is used. They were able to show that an improvement of approximately 50% in the S/N could be achieved if the temperature was increased from 30 to 140 °C at otherwise constant conditions, *e.g.* keeping the flow rate and the mobile phase composition constant. When using electrospray ionisation, the spraying and solvent evaporation is usually carried out without the introduction of a heated eluent. However, these processes are facilitated when a higher mobile phase temperature is applied. Care should be taken when interpreting these results, because a higher signal-to-noise ratio might also be observed if the flow behaviour of the mobile phase in the connecting tubing is enhanced, due to a plug-like flow profile in the transfer capillary. As Carr *et al.* have calculated, the band broadening in the post-column cooling unit is higher than in the preheating unit, based on the assumption that 20 cm of tubing is used for preheating and cooling.[59] In both cases, band broadening results because the viscosity of the mobile phase at the wall is different from the viscosity at the centre of the tubing. Usually, if a liquid is pumped through a tube, a parabolic flow profile results, because the velocity at the wall is lower than that in the middle of the tubing. By heating the wall, both the flows at the wall and at the centre are nearly identical and thus, preheating the mobile phase can have a positive effect on minimizing extra-column band broadening. It is different for the cooling at the column outlet when the effluent is cooled down after the column, a negative effect results because the viscosity near the walls is higher than that at the centre of the tubing. Thus, the band broadening is also more pronounced than in the preheating tube. While it is possible to focus the sample plug at the column head when solvent-gradient elution is applied and the gradient starts with a high water concentration, this is not possible after the sample bands leave the column. Therefore, a careful optimization of the high-temperature HPLC system as illustrated in Chapter 4 is of utmost importance to maximize the efficiency of the whole system. This is not only restricted to the column, but to all capillaries connecting the injector with the column, and the column with the detector. Depending on the type of detector, it should be evaluated whether a post-column cooling of the effluent is necessary, and if not, how to reduce the tubing as much as possible. I will return to this issue in Chapters 8 and 9, where you will see that the instrumental set-up often prohibits any progress in terms of separation and detection efficiency.

An experiment supporting these concepts has been performed by Pereira *et al.* for ESI and APCI-MS.[60] They carried out flow-injection analysis without a column, and they were able to show that a tremendous increase in peak area and peak height was observed when the eluent temperature was increased from 30 to 190 °C, with an optimum at approximately 150 °C. They concluded that these improvements are due to better analyte desolvation at higher temperatures and changes in solution chemistry in ESI or changes in gas-phase chemistry in APCI, which enhance the ionization process. However, higher eluent temperatures can also increase the noise due to column bleed, as I have shown in Chapter 5.1. Therefore, the authors also measured the S/N ratio for selected sulfonamides when a column was installed. While for APCI the S/N could be continuously improved on moving to higher temperatures, an optimum at around 120 °C was observed for ESI-MS.

It can be summarized that a careful optimization of the complete chromatographic system, including even the transfer capillary connecting the column to the detector, can help to improve the detection sensitivity. Temperature plays a pivotal role because it can support the nebulisation process of the above mentioned detectors. This means that cooling the eluent to ambient temperature is not necessary if an evaporative light-scattering detector, a charged aerosol detector or a mass spectrometer is used.

References

1. R. J. Maggs, *J. Chromatogr. Sci.*, 1969, **7**, 145.
2. J. Chmielowiec and H. Sawatzky, *J. Chromatogr. Sci.*, 1979, **17**, 245.
3. http://www.sim-gmbh.de/index.php?option = com_content&task = view&id = 64&Itemid = 502&lang = en (last accessed October 2009).
4. http://www.zirchrom.com/metalox.asp (last accessed October 2009).
5. http://www.selerity.com/main/main_products_hplc_9000.html (last accessed October 2009).
6. G. M. Schneider, *Phys. Chem. Chem. Phys.*, 2002, **4**, 845.
7. T. Teutenberg, P. Wagner and J. Gmehling, *J. Chromatogr., A*, 2009, **1216**, 6471.
8. P. L. Zhu, J. W. Dolan and L. R. Snyder, *J. Chromatogr., A*, 1996, **756**, 41.
9. L. R. Snyder, J. W. Dolan and J. R. Gant, *J. Chromatogr.*, 1979, **165**, 3.
10. J. W. Dolan, J. R. Gant and L. R. Snyder, *J. Chromatogr.*, 1979, **165**, 31.
11. J. W. Dolan, L. R. Snyder and M. A. Quarry, *Chromatographia*, 1987, **24**, 261.
12. J. W. Dolan, D. C. Lommen and L. R. Snyder, *J. Chromatogr.*, 1989, **485**, 91.
13. L. R. Snyder, J. W. Dolan and D. C. Lommen, *J. Chromatogr.*, 1989, **485**, 65.
14. P. L. Zhu, L. R. Snyder, J. W. Dolan, N. M. Djordjevic, D. W. Hill, L. C. Sander and T. J. Waeghe, *J. Chromatogr., A*, 1996, **756**, 21.
15. P. L. Zhu, J. W. Dolan, L. R. Snyder, D. W. Hill, L. Van Heukelem and T. J. Waeghe, *J. Chromatogr., A*, 1996, **756**, 51.
16. P. L. Zhu, J. W. Dolan, L. R. Snyder, N. M. Djordjevic, D. W. Hill, J. T. Lin, L. C. Sander and L. Van Heukelem, *J. Chromatogr., A*, 1996, **756**, 63.
17. J. W. Dolan, L. R. Snyder, N. M. Djordjevic, D. W. Hill, D. L. Saunders, L. Van Heukelem and T. J. Waeghe, *J. Chromatogr., A*, 1998, **803**, 1.
18. J. W. Dolan, L. R. Snyder, D. L. Saunders and L. Van Heukelem, *J. Chromatogr., A*, 1998, **803**, 33.
19. J. W. Dolan, L. R. Snyder, N. M. Djordjevic, D. W. Hill and T. J. Waeghe, *J. Chromatogr., A*, 1999, **857**, 1.
20. J. W. Dolan, L. R. Snyder, N. M. Djordjevic, D. W. Hill and T. J. Waeghe, *J. Chromatogr., A*, 1999, **857**, 21.

21. J. W. Dolan, L. R. Snyder, R. G. Wolcott, P. Haber, T. Baczek, R. Kaliszan and L. C. Sander, *J. Chromatogr., A*, 1999, **857**, 41.
22. L. R. Snyder and J. W. Dolan, *High-Performance Gradient Elution – The Practical Application of the Linear-Solvent-Strength Model*, Wiley-Interscience, Jon Wiley & Sons Inc, Hoboken, NJ, 2007.
23. H. Chen and C. Horvath, *Anal. Meth. Instr.*, 1993, **1**, 213.
24. C. B. Castells, L. G. Gagliardi, C. Rafols, M. Roses and E. Bosch, *J. Chromatogr., A*, 2004, **1042**, 23.
25. L. G. Gagliardi, C. B. Castells, C. Rafols, M. Roses and E. Bosch, *J. Chromatogr., A*, 2005, **1077**, 159.
26. C. B. Castells, C. Rafols, M. Roses and E. Bosch, *J. Chromatogr., A*, 2003, **1002**, 41.
27. S. M. C. Buckenmaier, D. V. McCalley and M. R. Euerby, *J. Chromatogr., A*, 2004, **1060**, 117.
28. F. Gritti and G. Guiochon, *J. Chromatogr., A*, 2006, **1131**, 151.
29. T. L. Chester and J. W. Coym, *J. Chromatogr., A*, 2003, **1003**, 101.
30. V. L. McGuffin, C. E. Evans and S. H. Chen, *J. Microcolumn Sep.*, 1993, **5**, 3.
31. A. M. Edge, S. Shillingford, C. Smith, R. Payne and I. D. Wilson, *J. Chromatogr., A*, 2006, **1132**, 206.
32. F. D. Antia and C. Horvath, *J. Chromatogr.*, 1988, **435**, 1.
33. J. Li and P. W. Carr, *Anal. Chem.*, 1997, **69**, 837.
34. Y. Yang, L. J. Lamm, P. He and T. Kondo, *J. Chromatogr. Sci.*, 2002, **40**, 107.
35. T. Kondo and Y. Yang, *Anal. Chim. Acta*, 2003, **494**, 157.
36. A. de Villiers, D. Cabooter, F. Lynen, G. Desmet and P. Sandra, *J. Chromatogr., A*, 2009, **1216**, 3270.
37. G. Guiochon, *J. Chromatogr., A*, 2006, **1126**, 6.
38. F. V. Warren and B. A. Bidlingmeyer, *Anal. Chem.*, 1988, **60**, 2821.
39. A. de Villiers, F. Lynen and P. Sandra, *J. Chromatogr., A*, 2009, **1216**, 3431.
40. D. Cabooter, F. Lestremau, A. de Villiers, K. Broeckhoven, F. Lynen, P. Sandra and G. Desmet, *J. Chromatogr., A*, 2009, **1216**, 3895.
41. S. Heinisch, G. Desmet, D. Clicq and J. L. Rocca, *J. Chromatogr., A*, 2008, **1203**, 124.
42. D. Cabooter, F. Lestremau, F. Lynen, P. Sandra and G. Desmet, *J. Chromatogr., A*, 2008, **1212**, 23.
43. D. Cabooter, J. Billen, H. Terryn, F. Lynen, P. Sandra and G. Desmet, *J. Chromatogr., A*, 2008, **1204**, 1.
44. F. Lestremau, A. de Villiers, F. Lynen, A. Cooper, R. Szucs and P. Sandra, *J. Chromatogr., A*, 2007, **1138**, 120.
45. D. Cabooter, S. Heinisch, J. L. Rocca, D. Clicq and G. Desmet, *J. Chromatogr., A*, 2007, **1143**, 121.
46. D. Clicq, S. Heinisch, J. L. Rocca, D. Cabooter, P. Gzil and G. Desmet, *J. Chromatogr., A*, 2007, **1146**, 193.

47. J. Billen, D. Guillarme, S. Rudaz, J. L. Veuthey, H. Ritchie, B. Grady and G. Desmet, *J. Chromatogr., A*, 2007, **1161**, 224.
48. S. Eeltink, P. Gzil, W. T. Kok, P. J. Schoenmakers and G. Desmet, *J. Chromatogr., A*, 2006, **1130**, 108.
49. G. Desmet, D. Clicq, D. T. Nguyen, D. Guillarme, S. Rudaz, J. L. Veuthey, N. Vervoort, G. Torok, D. Cabooter and P. Gzil, *Anal. Chem.*, 2006, **78**, 2150.
50. X. Wang, D. R. Stoll, P. W. Carr and P. J. Schoenmakers, *J. Chromatogr., A*, 2006, **1125**, 177.
51. P. W. Carr, X. L. Wang and D. R. Stoll, *Anal. Chem.*, 2009, **81**, 5342.
52. F. Gritti and G. Guiochon, *J. Chromatogr., A*, 2008, **1212**, 35.
53. F. Gritti and G. Guiochon, *J. Chromatogr., A*, 2008, **1178**, 79.
54. U. D. Neue, D. H. Marchand and L. R. Snyder, *J. Chromatogr., A*, 2006, **1111**, 32.
55. M. H. Chen and C. Horvath, *J. Chromatogr., A*, 1997, **788**, 51.
56. T. Gorecki, F. Lynen, R. Szucs and P. Sandra, *Anal. Chem.*, 2006, **78**, 3186.
57. A. de Villiers, T. Gorecki, F. Lynen, R. Szucs and P. Sandra, *J. Chromatogr., A*, 2007, **1161**, 183.
58. M. Albert, G. Cretier, D. Guillarme, S. Heinisch and J. L. Rocca, *J. Sep. Sci.*, 2005, **28**, 1803.
59. B. Yan, J. Zhao, J. S. Brown, J. Blackwell and P. W. Carr, *Anal. Chem.*, 2000, **72**, 1253.
60. L. Pereira, S. Aspey and H. Ritchie, *J. Sep. Sci.*, 2007, **30**, 1115.
61. T. Teutenberg, *Anal. Chim. Acta*, 2009, **643**, 1.
62. T. Teutenberg, H. J. Goetze, J. Tuerk, J. Ploeger, T. K. Kiffmeyer, K. G. Schmidt, W. G. Kohorst, T. Rohe, H. D. Jansen and H. Weber, *J. Chromatogr., A*, 2006, **1114**, 89.
63. S. Giegold, T. Teutenberg, J. Tuerk, T. Kiffmeyer and B. Wenclawiak, *J. Sep. Sci.*, 2008, **31**, 3497.

CHAPTER 7
Analyte Stability

The next serious issue which has to be considered when using high eluent temperatures is that a compound may undergo rapid degradation during the time it spends in the heated column. In this context, hydrolysis, oxidation, isomerisation and epimerization are the major types of chemical reactions that affect the stability of analytes and are considered undesirable. The fear of such reactions has contributed significantly to the widespread reluctance to use high-temperature HPLC by the pharmaceutical industry.

There is a concept which describes the processes involved in a chromatographic column concerning analyte stability. According to Melander, Antia and Horváth, on-column reactions are conveniently characterized by the dimensionless Damköhler number (Da), which may be defined as the ratio of the residence time of a compound in the column to the relaxation time for the on-column reaction.[1,2] When there is no reaction ($Da \ll 1$) or the reaction is very fast ($Da > 50$), single peaks are expected. Excessive peak broadening and irregularly shaped elution profiles are observed at intermediate values of the Damköhler number ($0.1 < Da < 50$). In this case, the analyte decomposes over the time scale of the separation, and the intermediates and fragments will differentially migrate over the length of the column to produce a broad bimodal elution profile. All things being equal, the Damköhler number is expected to decrease as the column length decreases, the linear velocity increases, the retention factor decreases and when the rate of the reaction is small. The authors further state that in analytical chromatography, it is of little consequence whether the reaction has proceeded almost to completion (as indicated by large Da values), or not at all (as suggested by $Da \ll 1$). In the first case, the analyte is converted into a single product, whereas in the second case the analyte remains unchanged. Although the authors comment that analytical efficiency is not compromised in either case, the full conversion of a compound may be devastating during the drug development process in the pharmaceutical industry. Clearly the user needs to distinguish between analytical separations (where the requirement is simply to correlate the signal to an original concentration) and preparative separations (where isolation of a target molecule is required). While in the first instance complete degradation has no consequences (as long as the process is reproducible and a proper quantification can be obtained), in the second instance degradation is not acceptable.

In the following paragraph, I would like to outline a practical approach to gaining an overview as to whether the compounds in a sample mixture are degraded as the temperature is increased.

7.1 Evaluation of Analyte Stability using UV Detection

A very easy approach to evaluating analyte stability is to use a UV detector at a fixed wavelength. In liquid chromatography, the peak areas at a fixed wavelength should be constant provided that the flow rate is not changed. More specifically, the equation for peak area for a concentration-dependent detector can be written as:

$$A_\phi = \frac{\varepsilon \cdot b \cdot W_a}{F} \quad (7.1)$$

where the subscript Φ refers to a specific eluent; ε is the molar absorptivity for the analyte; b is the flow cell path length; W_a represents the total mass of the analyte; and F is the flow rate of the mobile phase.

Therefore, a simple experiment can be performed where the dependence of the peak areas of the target analytes on temperature are measured. As the temperature is increased, the retention factors usually decrease, which means that the time the analyte spends on the hot column is reduced (please remember the van't Hoff equation which describes the temperature-dependent behaviour of retention, see Chapter 6.4.1). If all other parameters are kept constant, the peak area of the compounds should not change. A plot of peak area against temperature for all analytes will reveal if there is a major deviation from this rule. In Figure 7.1, such a plot is shown as an example for the steroid mixture I introduced in Chapter 6.5.1. It can be clearly seen that the peak areas remain constant over the given temperature range. This means that a degradation of the selected steroids was not observed.

However, a thermally induced degradation may already occur at a much lower temperature, which is highlighted in Figure 7.2. Here, the same experiment was carried out using different model compounds.

It is obvious that amoxicillin already starts to degrade at 40 °C, while the other compounds are stable over the selected temperature range. This means that it is not possible to define a threshold temperature above which a thermal degradation will occur. It clearly depends on the analyte and the chromatographic conditions, as I will illustrate in the following paragraphs.

In addition, besides the peak area, the peak itself might reveal if a thermally induced degradation had occurred. This is highlighted in the next experiment, in which sulfathiazole and N^4-acetylsulfamethazine were eluted from a

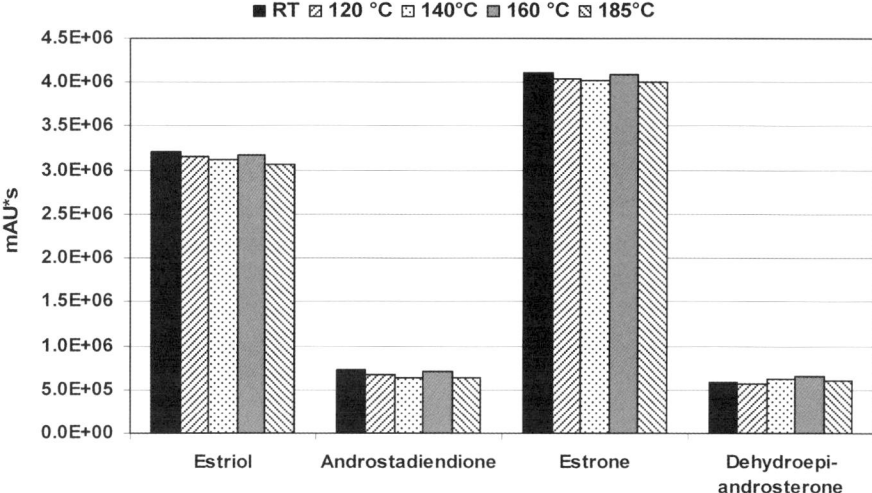

Figure 7.1 Comparison of the peak areas of selected steroids at different temperatures. Chromatographic conditions: mobile phase at 25 °C: water–acetonitrile (75 : 25 v/v); mobile phase at 120, 140, 160 and 185 °C: deionized water; column: ZirChrom-PBD (4.6 × 150 mm ID; 3 μm, 300 Å); flow rate: 1 ml min^{-1}.

carbon-clad zirconium dioxide stationary phase. The temperature was increased from 60 °C to 180 °C in increments of 20 °C (see Figure 7.3). For sulfathiazole, the peak height increased with increasing temperature (see Figure 7.3a). This means that the broad elution profile which resulted at a relatively low temperature becomes narrower and thus, a better signal-to-noise ratio was obtained. For quantitative analysis, this is a clear advantage.

In Figure 7.3b, the elution of N^4-acetylsulfamethazine is shown. Here, a completely different elution profile was observed as the temperature is increased from 60 °C to 180 °C. While the peak profile given in the first chromatogram in Figure 7.3b looked quite normal, increasing the temperature from 80 °C to 120 °C shows that the peak nearly disappeared. In effect, only a very broad peak was obtained at 120 °C, which is a sign that an on-column reaction has taken place. When the temperature was further increased, a peak could again be detected at 160 °C. Nevertheless, the peak was very broad and showed a small shoulder at the leading edge with strong tailing. This indicates that there were probably two or more species and that this was a dynamical equilibrium. Quite surprisingly, at the highest temperature employed (180 °C), a sharp peak was again obtained. However, there was also a small leading peak in front of the large peak. Between these two peaks, there is a reaction zone because both peaks are connected *via* a plateau. This means that in contrast to sulfathiazole, an on-column reaction takes place for N^4-acetylsulfamethazine under high-temperature conditions.

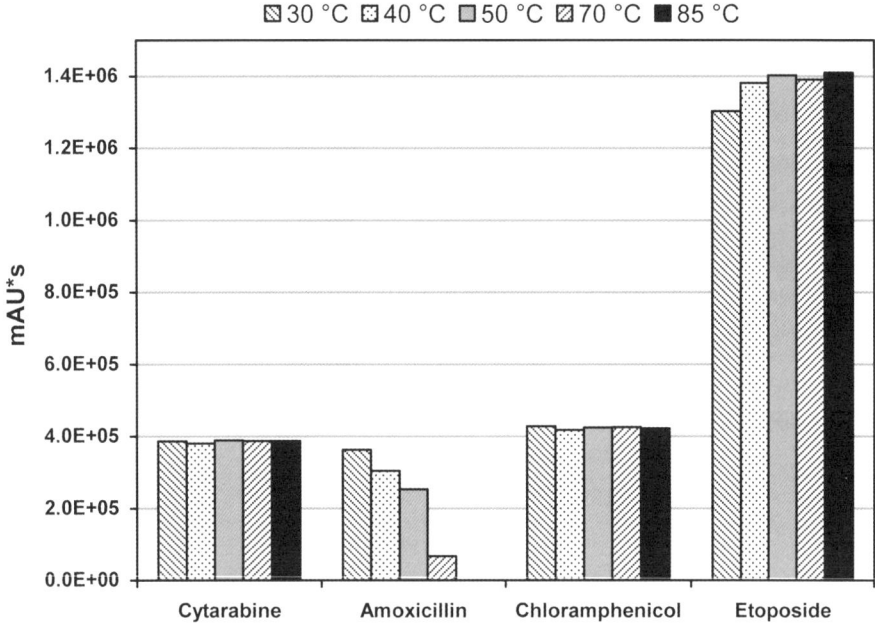

Figure 7.2 Comparison of the peak areas of selected cytostatic and antibiotic drugs at different temperatures.[6] Chromatographic conditions: column: ZirChrom-PBD (4.6 × 150 mm ID; 3 μm, 300 Å); mobile phase: water containing 0.1% formic acid–acetonitrile (85 : 15 v/v); flow rate: 1 ml min^{-1}. (Reproduced with kind permission from Elsevier.)

These two examples highlight that besides a peak area measurement, a visual inspection of the elution profile is necessary to evaluate if there is an on-column reaction of the selected species.

7.2 Influence of the Stationary Phase on Analyte Stability

Temperature is not the only parameter which can affect a degradation of sample components. The pH of the mobile phase can also have a detrimental effect on analyte stability. What might be even more surprising is that the stationary phase itself can accelerate degradation. A very interesting phenomenon was observed by Giegold *et al.*, who investigated the stability of an active pharmaceutical ingredient on two different stationary phases. One of these phases was a polystyrene–divinylbenzene column, while the other packing consisted of carbon-clad zirconium dioxide. The temperature was raised in increments of 20 °C and the peak areas were recorded. Figures 7.4 and 7.5 show the changes in the chromatograms obtained for the elution of thalidomide. It can be clearly seen that a rapid breakdown of the substance occurred if the

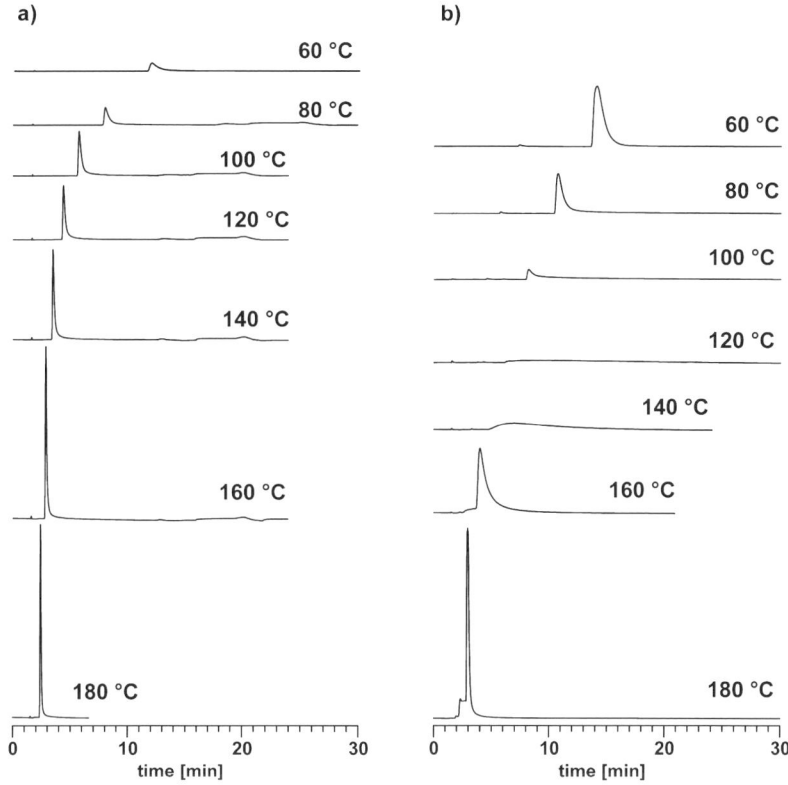

Figure 7.3 a) Elution of sulfathiazole on a ZirChrom Carb column (150 × 4.6 mm ID: 3 μm) at: 60 °C, 80 °C, 100 °C, 120 °C, 140 °C, 160 °C and 180 °C. Chromatographic conditions: flow rate: 1 ml min^{-1}; mobile phase: deionized water–acetonitrile , each containing 0.1% formic acid (180 : 20 v/v); detection: UV-DAD at 270 nm. b) Elution of N^4-Acetylsulfamethazine on a ZirChrom Carb column (150 × 4.6 mm ID; 3 μm) at 60 °C, 80 °C, 100 °C, 120 °C, 140 °C, 160 °C and 180 °C. Chromatographic conditions: flow rate: 1 ml min^{-1}; mobile phase: deionized water–acetonitrile, each containing 0.1% formic acid (70 : 30 v/v); detection: DAD at 270 nm.

carbon-clad metal oxide column was used and the temperature was above 100 °C. In contrast to this, there was no degradation on the polymeric stationary phase.

In addition, the peak height of the thalidomide peak increased at higher temperatures on the polymeric column, improving the signal-to-noise ratio significantly. Hence, a better quantification is possible which was also supported by a more Gaussian peak shape. The asymmetry was reduced from 2.01 at 60 °C, to 1.38 at 180 °C, which means that the peak eluted as a more symmetrical band. This example quite nicely underlines that even the stationary phase can have a pronounced impact on analyte stability. Catalytic effects on the zirconia surface might have contributed to this degradation.

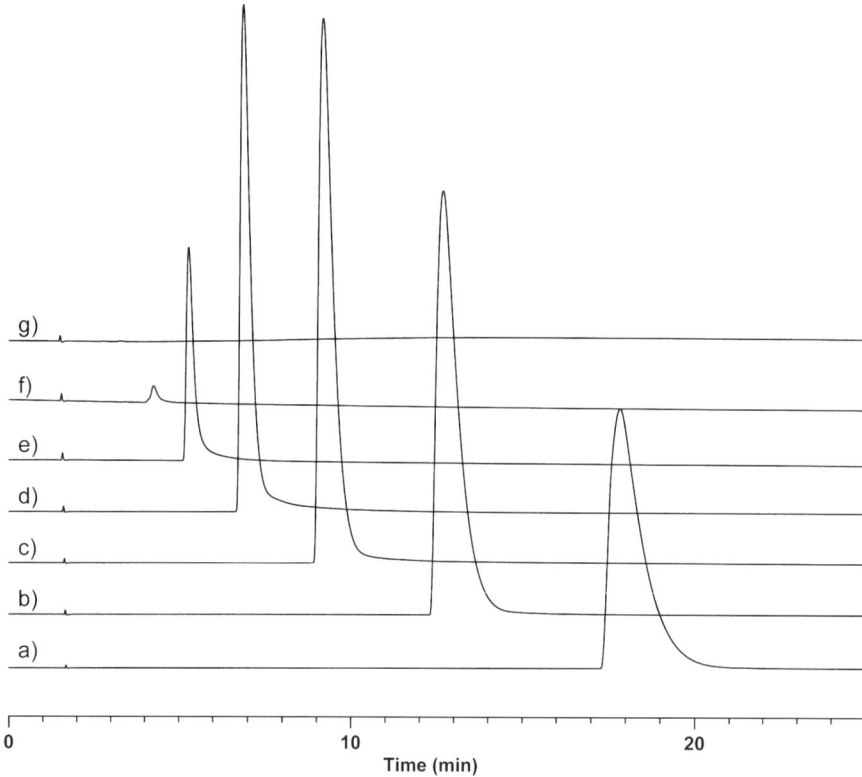

Figure 7.4 Monitoring of the on-column degradation of thalidomide on a ZirChrom Carb column (150 mm × 4.6 mm ID; 3 μm) from (a) 60 °C to (g) 180 °C in 20 °C increments. Chromatographic conditions: flow rate: 1 ml min^{-1}; mobile phase: deionised water–acetonitrile, each containing 0.1% formic acid (75 : 25); detection: DAD at 300 nm.[7] (Reproduced with kind permission from Elsevier.)

The most important conclusion which can be drawn from this example is that analyte stability should always be determined on the column which is used to run the method. Do not use or select a reference column to measure analyte stability, and then transfer the method to another column. The result may be that the analyte is stable on one column but will degrade on another column under identical conditions.

7.3 Definition of Critical Criteria for Analyte Stability

In order to better evaluate if a particular analyte should be rejected for high-temperature HPLC, Carr and Thompson proposed some criteria which may help to identify the problems related to analyte degradation.[3] Since I find these criteria very useful, I have noted these statements here and have made some

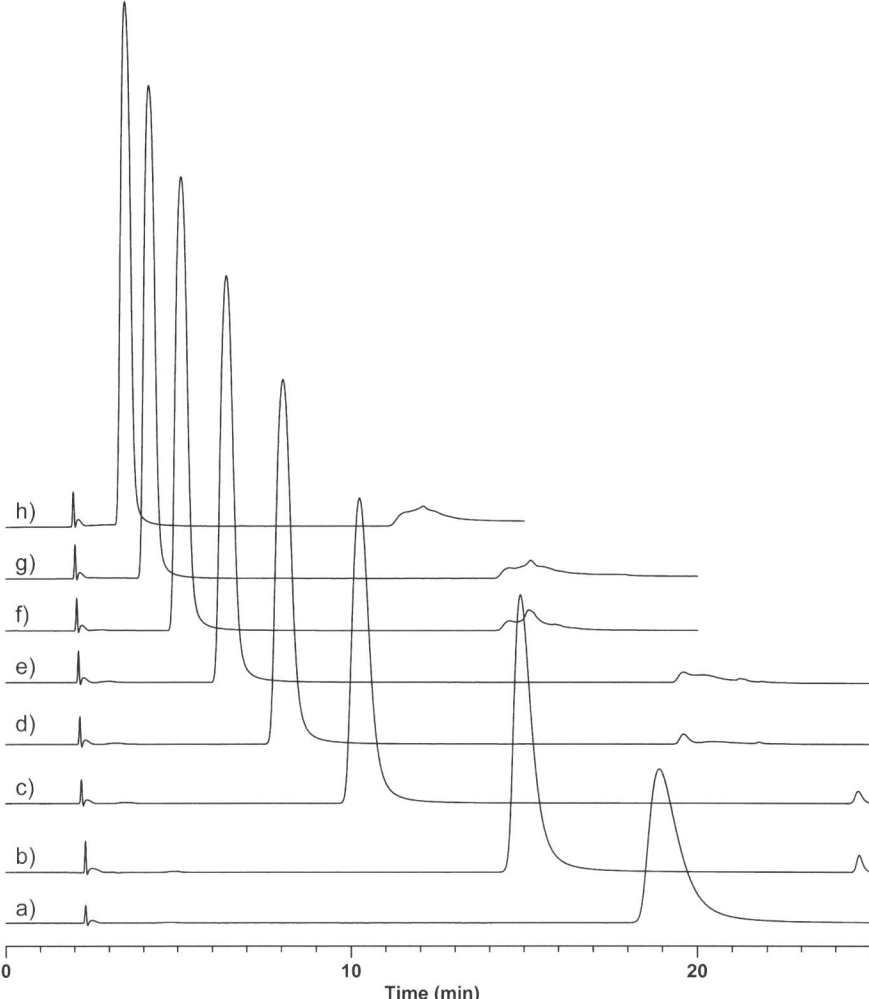

Figure 7.5 Monitoring of the on-column degradation of thalidomide on a PLRP-S column (150 mm × 4.6 mm ID; 3 μm) from (a) 40 °C to (h) 180 °C in 20 °C increments. Chromatographic conditions: flow rate: 0.2 ml min^{-1}; mobile phase: deionised water–acetonitrile, each containing 0.1% formic acid (80 : 20); detection: DAD at 300 nm.[7] (Reproduced with kind permission from Elsevier.)

comments on how to overcome difficulties due to a poor analyte stability. On changing the temperature:

1. Is the reliability of the analytical calibration curve decreased?
2. Is a significant intercept induced in the calibration curve?
3. Is the sensitivity significantly diminished?

4. Is the peak shape severely distorted?
5. Do new peaks show up that interfere with the quantification of the analyte or impurities?
6. Does a high-temperature eluent cause chemical reactions that alter the concentration, or produce products that would interfere with the quantification of impurities or other constituents in the sample?

The authors came to the conclusion that if the answers to all these questions for a particular analyte are "no", then a possible on-column reaction was not important and quantification at high-temperature ought not be problematic. It is clear that even in the absence of any on-column reaction, the sensitivity of the method might be compromised, because of an increased column bleed at elevated temperature. This might be the case for mass spectrometric detection where the signal also strongly depends on the "matrix" which co-elutes with the analyte. Usually, target compounds can be detected selectively, depending on the mode of detection and the instrument which is used. However, the fact that only one specific mass is detected does not mean that other compounds which co-elute with the analyte of interest will not have a positive or negative impact on the ionization efficiency of that compound. Hence, column bleed might be more problematic when mass spectrometric detection is used (see Chapter 5.1 for more information on column bleed).

As was also evident from the examples in section 7.1, a distorted peak shape can make an accurate quantification impossible. However, what needs to be considered is that the occurrence of additional peaks at high eluent temperatures does not necessarily mean that the target compounds (or the API) have been degraded on-column. Often, other peaks appear because the selectivity of the phase system is different when the temperature is changed. I think this phenomenon is often misinterpreted, because a two-dimensional variation of operating conditions is very powerful for finding the best separation conditions. In this context, temperature plays an important role in method development, as I have shown in the previous chapter.

Nevertheless, there is no general rule or statement which can be made about analyte stability at elevated temperatures. I think the rules which were proposed by Thompson and Carr are quite helpful for evaluating whether there may be any problems with regards to analyte degradation. However, the analytical scientist has to use his or her expertise to correctly interpret the results.

With new technologies at hand, fast separations in the range of a few seconds can be easily obtained. In this respect, one of the advantages of high-temperature HPLC is to speed up separation, as was demonstrated in Chapter 4. Very large gain factors can be obtained, so that the time the analyte spends in the heated zone is brought to a minimum. As was also demonstrated by Horváth, the separation of large macromolecules, like proteins and peptides, can be achieved at elevated temperatures, just by increasing the speed of the separation.[4] Therefore, it can be concluded that the fear of a rapid and inevitable degradation of sample constituents at high eluent temperature is, in most cases, overstated. It should be noted that pH can also have the same effect on analyte stability as high eluent temperatures. However, chromatography at pH 2 or 10 is not considered a problem as long as the column is stable. Therefore, the practitioner should not be afraid to use high eluent temperatures.

A very nice example, which serves to illustrate how method development should proceed in cases where thermally labile compounds are analyzed, is given by de Villiers.[5] The paper showed that an absolute increase in column efficiency at higher temperatures was only achieved, when the temperature reduced the effect of secondary equilibria. The authors noted that at higher temperatures, an absolute increase in column efficiency was observed. This was due to an inter-conversion reaction between two species, which increased at higher temperatures, and thus peak broadening was reduced. This is a phenomenon which relates to a secondary equilibrium and improves the absolute column efficiency, which is observed in the decrease in the minimum plate height. However, for the analysis of the substances studied by de Villiers it is important to consider that a thermally induced degradation could also have taken place. Although higher eluent temperatures favour an increase in the efficiency of the separation, it is also possible for thermally induced on-column degradation to occur. This has been shown by calculating the temperature-dependence of the Damköhler number. Thus, the best compromise was achieved when the analysis was carried out at 50 °C. Under these conditions, no degradation can be expected for an analysis time of two hours and efficiency could be increased due to a reduction in the minimum plate height.

References

1. W. R. Melander, H. J. Lin, J. Jacobson and C. Horvath, *J. Phys. Chem.*, 1984, **88**, 4527.
2. F. D. Antia and C. Horvath, *J. Chromatogr.*, 1988, **435**, 1.
3. J. D. Thompson and P. W. Carr, *Anal. Chem.*, 2002, **74**, 1017.
4. H. Chen and C. Horvath, *Anal. Meth. Instr.*, 1993, **1**, 213.
5. A. de Villiers, D. Cabooter, F. Lynen, G. Desmet and P. Sandra, *J. Chromatogr., A*, 2009, **1216**, 3270.
6. T. Teutenberg, *Anal. Chim. Acta*, 2009, **643**, 1.
7. S. Giegold, M. Holzhauser, T. Kiffmeyer, J. Tuerk, T. Teutenberg, M. Rosenhagen, D. Hennies, T. Hoppe-Tichy and B. Wenclawiak, *J. Pharm. Biomed. Anal.*, 2008, **46**, 625.

CHAPTER 8
Special Hyphenation Techniques

Although I have shown in Chapter 4 that all binary solvent systems can be used at high temperatures, the huge potential of high-temperature liquid chromatography lies in the fact that it allows special hyphenation techniques to be employed. Nearly all of these techniques, which will be introduced in the following sections, rely on a mobile phase consisting of pure water, or water with only a minimal portion of an organic co-solvent. Here, the effect is exploited that the static permittivity of water, or a binary mixture of water and an organic co-solvent, significantly decreases if the temperature is increased.[1]

In Chapter 4.3 it was clearly shown that the static permittivity of water drops from 80 at ambient temperature, to approximately 35 at 200 °C (see Figure 4.13). This means that the characteristic properties of water become similar to non-polar solvents, like methanol or acetonitrile. However, a direct comparison of the water's polarity with the polarity of other solvents is not possible, because the elution strength of a solvent cannot be reduced to a single parameter. Nevertheless, the elution strength of water increases with higher temperature. This means that it is possible to completely reduce the organic solvent, as I have already demonstrated in Chapter 6.5. What does this mean for method development? If an organic co-solvent cannot be used to change the elution strength of the mobile phase, as is commonly done in classical reversed-phase HPLC, the only option to elute a mixture of compounds with a wide range of polarities is to use temperature programming. The technical requirements of the heating system and the column have been given in Chapters 3 and 5, respectively. Also, the principal strategy for method development by applying temperature programming has been explained in Chapter 6. If a carbon-containing co-solvent cannot be used, retention and selectivity both need to be influenced by temperature. This means that you have not the full arsenal at hand to optimize chromatographic resolution and hence selectivity. This is in fact a drawback, but as was demonstrated in Chapter 6, the replacement of the binary eluent, and thus the organic co-solvent, by a pure water mobile phase is possible.

What has to be considered is that there are only a few options to change the selectivity during method development when working with pure water. First of all, the number of available columns which have a long-term stability at high temperatures and can be used with a pure water mobile phase is currently quite limited.[2,3] Also, the selectivity cannot be influenced by different organic co-solvents, because the use of these modifiers is in most cases strictly forbidden. The only remaining parameter besides temperature which can be included for method development is the pH of the mobile phase. Although not frequently used in everyday method development, pH is a very powerful parameter for adjusting the selectivity and for optimizing resolution.

However, in the last few years, new hardware has become commercially available, which includes dedicated heating ovens, as well as stationary phases for high-temperature HPLC. It is no longer a dream that these techniques, which I will explain in the following sections, can be put into practice.

8.1 Flame Ionization Detection

Probably one of the best known techniques directly related to superheated water chromatography is the hyphenation of high-temperature HPLC with flame ionization detection.[4–10] What was the primary driving force to establish this technique? In a routine separation laboratory, a flame ionization detector is only used in combination with gas chromatography. The reason is that the mobile phase, which might consist of hydrogen, helium or nitrogen, is mixed with hydrogen and completely burned in air at the tip of a jet. During this process, organic compounds are ionized and then detected. A flame ionization detector is also often termed a "universal detector" because it is said that there exists a linear relationship between the signal intensity and the amount of carbon in the eluent. Therefore, at least theoretically, the amount of carbon in an unknown compound can be determined with an external calibration standard. Furthermore, the FID is thought to be a good alternative to a UV detector, when analytes lacking a chromophore (such as lipids) need to be detected. This means that there is a high motivation to apply this concept in liquid chromatography too. The only problem is the need to get rid of the organic solvent in the mobile phase. So what works well for gas chromatography, where the carrier gas is inorganic, does not easily transfer to liquid chromatography, where a binary eluent mixture of water and an organic co-solvent is most commonly used.

So what are the implications if you would like to use a flame ionization detector for liquid chromatography? In this case, the conventional mobile phase always contains organic solvents, which will be burned, making the detector blind to the compounds to be detected. Early technical solutions were based on interfaces leading either to eluent evaporation from a wire, disc or belt, leaving the residual analyte to be detected. Although this approach allows for a use of binary solvent systems including organic co-solvents, such a system had two major drawbacks, namely the complexity of the design and the very high probability of vaporizing the sample with the solvent. Various technical improvements included a drop headspace interface which conveyed volatile analytes from the eluent to the FID, or an eluent-jet interface which generated a jet of droplets by inducing a sharp temperature gradient at the tip of the introduction capillary.

An alternative way to achieve HPLC-FID coupling involves the use of neat water as the mobile phase. In this case, there is no organic solvent to be removed before the mobile phase is introduced into the detector through a capillary restrictor. Numerous parameters, including hydrogen and air flow rates are known to have a significant effect on both FID sensitivity and FID noise. Optimized values leading to the best performance in terms of signal-to-noise ratio were found for hydrogen and air flow rates, with various water flow rates ranging from 20 to 200 µl min^{-1}. In all these different approaches, the restrictor outlet was placed 3 to 5 cm below the tip of the FID flame jet, in order to keep the FID lit. *The conclusions which can be drawn from these various studies suggest that firstly, the optimized hydrogen and air flow rates depend on the water flow rate; secondly, the best FID response is reached with a hydrogen flow rate as low as possible to keep the flame lit; and thirdly the best performance is obtained at a very high detector temperature of about 300 to 350 °C.*

Before I discuss some applications, I will focus on some instrumental considerations which are important for obtaining a stable signal. As was shown in Chapter 4.1.2, when the temperature of the mobile phase is increased so that a phase transition is likely to occur, a solvent perturbation can lead to split or distorted peaks (see Figure 4.7). It should be assumed that the same is true when an HPLC-FID system is designed, where the transfer capillary leading from the column outlet to the inlet of the FID is placed in the same oven as the column. Wu and co-workers described a system where neat water at elevated temperatures was used for the elution and detection of selected phenols and alcohols.[11] They used a 10 cm × 100 µm ID stainless-steel tube to connect the column with the FID, and estimated that the water exists in the vapour state in less than 10% of the column length. As the mobile phase was converted to a gas near the outlet of the column when the temperature was above 150 °C, a more stable signal was obtained when compared with that at a lower temperature of elution. This observation is quite the opposite to the effects I have described in Chapter 4.1, where a phase transition of the mobile phase in the column led to an immediate and irreversible failure of the stationary phase. The results reported by Ingelse and co-workers clearly confirm the effects which are described in Chapter 4.1.2 of this monograph. The authors noted that even if the column was operated isothermally, the use of linear restrictors occasionally resulted in detector spiking due to evaporation and subsequent condensation of the eluent in the restrictor, which they termed "sputtering".[12] This became even worse when temperature programming was applied. The use of tapered restrictors, as frequently used in supercritical fluid chromatography, helped to eliminate the problem of solvent sputtering, but suffered from easy blockage. Therefore, they only used linear restrictors, but placed the transfer capillary in a second oven which was kept at a constant temperature below the boiling point of water. These two examples clearly demonstrate that the system needs to be carefully designed. The question as to whether the restrictor has to be placed outside the column oven also depends on the method. Provided that the temperature is not increased above 100 °C, corresponding to the normal boiling point of water, there is no need to worry about a phase transition, solvent perturbation or sputtering effect. However, if a large temperature interval is considered and the temperature needs to be increased up to 200 or even 250 °C, a set-up as described by Ingelse and co-workers seems to be very well suited to avoid such problems. If temperature programming is used it should be noted that starting with a low temperature might lead to an excessive back pressure if a very small ID capillary is used as a linear restrictor. Hence, the approach described by Ingelse and co-workers to use a transfer line, which is heated independently from the column, will guarantee that neither an excessive back pressure is observed at a low column temperature, nor is there a problem related to a phase transition of the water. With this set-up, a constant back pressure was obtained regardless of the temperature programme.

In fact, various authors very successfully employed flame ionization detection with a water only mobile phase. I would like to give some selected examples and show the respective chromatograms. Wu and co-workers used superheated water to separate mixtures of phenols and alcohols. The authors used a stationary phase based on polybutadiene-coated zirconium dioxide, which they modified by crosslinking the PBD-coated zirconia particles to a polyoctylmethylsiloxane-coated capillary.[11] The resulting chromatograms are shown in Figure 8.1, exhibiting a very nice baseline separation of the respective test mixtures.

Using a capillary column is a clear advantage over conventional size columns with this application. The reason is that at flow rates above 200 μl min^{-1}, which are directly introduced to the FID, the flame will be extinguished immediately. Therefore, the internal diameter of the column should be adjusted to the flow rate which can be introduced into the detector. Furthermore, it should be considered that at higher temperatures, the flow rate needs to be increased in order to make sure that the column is operated near the van Deemter minimum. Too low a flow rate will inevitably lead to a loss in separation efficiency, and will add to excessive band broadening when the column is operated in the B-term dominated region. In this case, a flow splitter has to be used when the separation needs to be carried out at high flow rate, but then only a fraction of the sample is introduced into the detector. This application highlights the problem which results if the method can only be run in isocratic mode, because a huge loss in efficiency is observed when the column is not operated in its respective van Deemter minimum. A band compression as in solvent gradient elution is not possible and hence, the column diameter has to be adjusted depending on the maximum temperature and flow rate. Otherwise, broad peaks will be observed due to longitudinal diffusion. However, the sensitivity of the method has to be considered, because a certain sample amount needs to be injected on-column to produce a signal. Otherwise, the column is operated under optimum conditions, but the method fails because of a poor limit of detection.

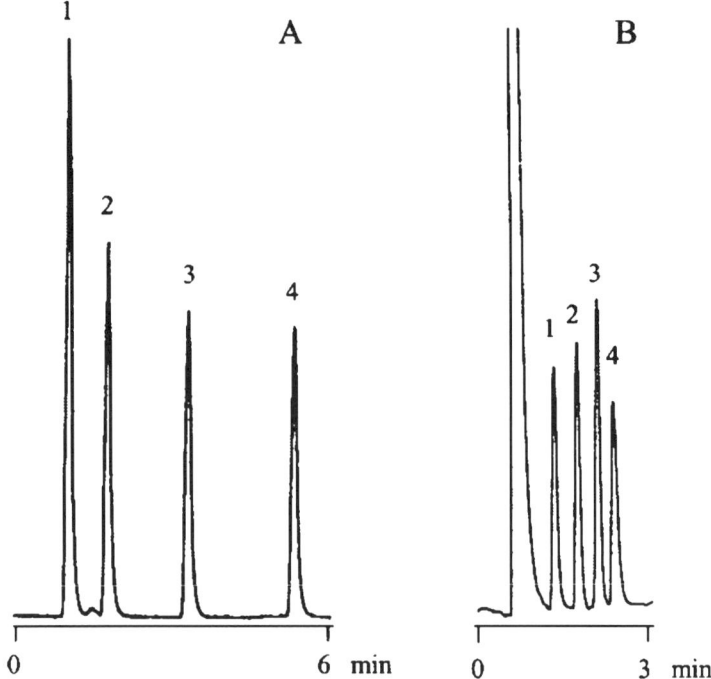

Figure 8.1 Chromatograms of (A) alcohols and (B) phenols. Chromatographic conditions for (A) alcohols: column: continuous-bed capillary (25.2 cm × 250 μm ID); pressure: 200 atm; temp. gradient: 125 to 165 °C at 5 °C min^{-1}. Peaks: 1, methanol; 2, isopropanol; 3, 3-methylbutanol; and 4, 1-heptanol. Chromatographic conditions for (B) phenols: column: continuous-bed capillary (33 cm × 250 μm ID); pressure: 250 atm; temp. gradient: 165 to 200 °C at 10 °C min^{-1}. Peaks: 1, 4-chlorophenol; 2, 3-methyl-4-chlorophenol; 3, 2,4,6-trimethylphenol; and 4, 2,4-dichlorophenol.[11] (Reproduced with kind permission from John Wiley & Sons, Inc.)

The separation of alcohols has been described by many groups using high-temperature HPLC with flame ionization detection and a pure water mobile phase. Guillarme and co-workers developed a fast method on a PBD-coated zirconium dioxide column to separate five linear alcohols.[9] The separation was optimized by increasing the temperature from 30 to 120 °C, resulting in a runtime below 1.5 minutes which is depicted in Figure 8.2. A high sample throughput can therefore be achieved with this set-up. This example also highlights that a PBD-coated zirconium dioxide column is very well suited to this kind of application.

However, taking all these examples into account, there are two issues which are rarely addressed. First of all, most research groups use linear alcohols as test mixtures. While these compounds are difficult to detect with a UV detector, a flame ionization detector is ideally suited for this kind of application since the detection is not limited to analytes with a chromophoric

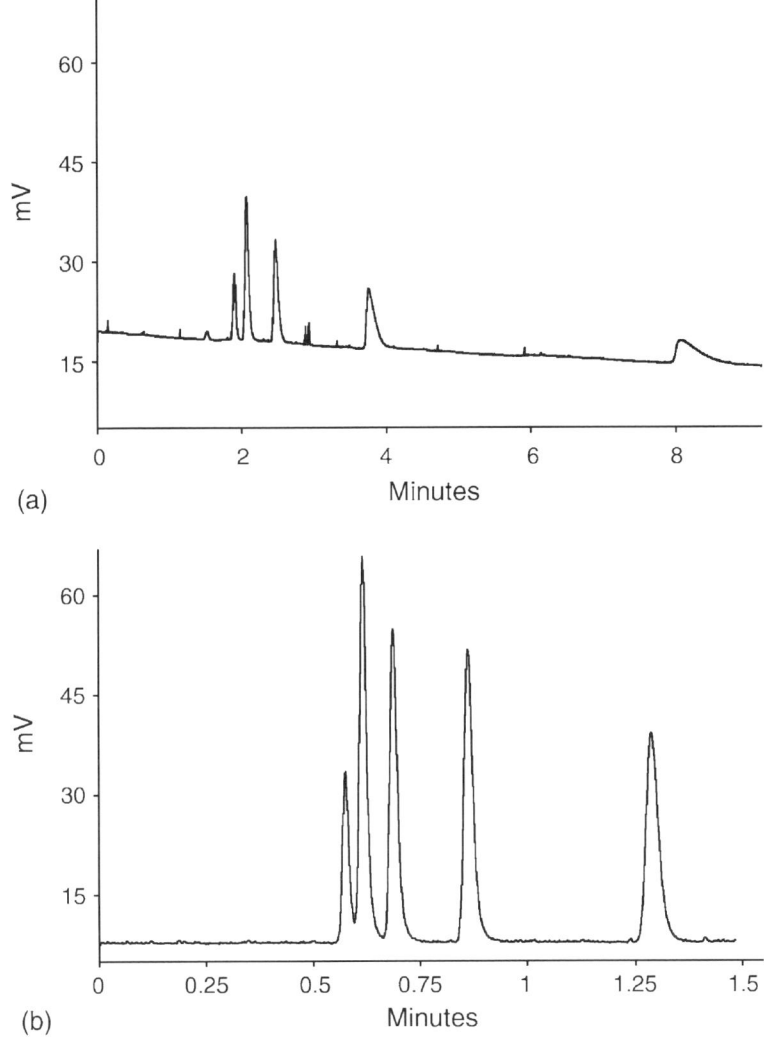

Figure 8.2 Separation of a mixture of five linear alcohols (150 ng injected) on a Zir-Chrom PBD column (100 mm × 0.5 mm ID) with a mobile phase of water at (a) 30 °C and 10 μl min^{-1}, and (b) at 120 °C and 30 μl min^{-1}. Elution order: methanol, 1-propanol, 1-butanol, 1-pentanol and 1-hexanol.[9] (Reproduced with kind permission from Elsevier).

system. The problem is that if "real" samples have to be analyzed, there may be a large number of other compounds which form the so-called "matrix" and which may easily interfere with the target analytes, so that co-elution becomes inevitable. This also means that a precise quantification of target substances is not possible and hence, sample clean-up has to be applied prior to HPLC analysis.

The second problem is that the matrix may consist of compounds which are highly diverse in their polarity or hydrophobicity. This means that non-polar compounds can accumulate on the column and will be washed off after a certain time, which will lead to column bleed and an increased signal-to-noise level.[i] This problem occurs when water only is used as the mobile phase in the analysis of complex samples. Hence, the column needs to be cleaned periodically with a strong organic solvent in order to elute these strongly bound compounds from the stationary phase. The reason for this is that the elution strength of water, even at very high temperatures, may be too low to remove these hydrophobic compounds. Unfortunately, in many experimental protocols for real samples, details regarding column cleaning procedures are absent. While a column cleaning procedure may be easily incorporated using a switching valve, the difficulty of analyzing real samples, possibly containing hundreds of compounds, has not been addressed.

Another problem is the sensitivity of the method. Usually, the main components of the sample can be easily detected and quantified. This has been demonstrated by Yarita and co-workers who used HPLC-FID to determine ethanol in various alcoholic beverages.[7] The technique was not suitable for trace analysis, however, because the limit of detection was rather poor. Therefore, enrichment techniques, in combination with sample clean-up, are required if minor components are to be analyzed.

It can be summarized that the coupling of HPLC with flame ionization detection is only possible when pure water is used as a mobile phase. The addition of organic co-solvents in the mobile phase presents a problem, because a reliable measurement of organic compounds cannot be achieved. Since this is an isocratic technique, the column diameter and the flow rate should be carefully selected so that the column is run in its respective van Deemter minimum. Capillary columns are ideally suited because the flow rate which is used during introduction into the detector can be kept low, so that the flame is not extinguished. However, in order to analyze complex samples, the stationary phase needs to be eluted with a strong organic solvent to remove non-polar compounds from the column.

8.2 LC-NMR

Nuclear magnetic resonance (NMR) spectroscopy has a significant advantage over many other analytical methods in that a vast amount of structural information can be gained in a single analysis, whilst conserving the sample for subsequent interrogation by other techniques. Although NMR is less sensitive than several other popular structure-elucidation techniques, such as, for example, mass spectrometry, the information obtained from the NMR spectrum can be sufficient to identify an unknown analyte.

---

[i] Please note that in this case, column bleed arises because of a gradual wash-off of strongly bound compounds from the column. In contrast to this, column bleed may be also encountered due to a thermally induced degradation of the column, as described in Chapter 5.1.

Generally, HPLC-NMR analysis can be configured using two different modes of operation, called "on-flow" or "stop-flow", to detect the separated analytes. During an "on-flow" experiment, a series of NMR spectra is acquired rapidly as the HPLC eluent flows through the NMR probe. "On-flow" analysis is ideal when investigating concentrated samples for which NMR spectra can be measured without extensive signal averaging. However, most samples are not sufficiently concentrated to permit analysis by on-flow experiments. In addition, lengthy 2D NMR experiments, which provide the detailed information about spin connectivity required for complete structure elucidation, cannot be acquired in the limited time available for on-flow experiments. Therefore, stop-flow experiments, in which the analyte is held in the NMR detection probe, are used for extensive signal averaging or for the acquisition of 2D NMR spectra.

Stop-flow NMR experiments can be configured in several ways. The simplest approach halts the HPLC pump at the appropriate time, stopping the flow of mobile phase and trapping the peak of interest in the NMR probe. Although this approach is simple and effective, once the HPLC pump is stopped, band broadening processes compromise the separation of peaks remaining in the column. Therefore, the stop-flow approach is mainly used to detect a single chromatographic peak from each sample injection, presenting a problem for the analysis of mass-limited samples. In an alternative approach, different fractions from the HPLC effluent are trapped into one of several sample-collection loops. Many peaks can then be collected from a single chromatographic run into a series of sample loops and transferred one at a time to the NMR. A related strategy completely decouples the separation from the NMR detection by trapping chromatographic peaks on solid-phase extraction cartridges, which can be subsequently eluted with deuterated solvents for NMR analysis.

Which approach is most suitable for high-temperature HPLC? If possible, on-flow elution should be selected when the concentration is high enough for an NMR spectrum to be obtained. If this is not possible, stop-flow experiments have to be carried out with the inherent drawback that excessive peak broadening will occur if the flow is halted. This effect is even worse at high eluent temperatures, since, as was shown in Chapter 6.4.3, the optimum in the van Deemter curve shifts to higher linear velocities at higher temperatures. Therefore, peak broadening will be aggravated at high eluent temperatures under stop-flow conditions. In that respect, peak trapping would be a good option. However, the enrichment of the fractions on solid-phase cartridges would require the use of deuterated organic solvents, which would render the advantage in eliminating the organic co-solvent in the separation step useless. Therefore, sample storage in different loops and analysis of these fractions one at a time should be the most promising approach.

The hyphenation of HPLC with NMR using deuterated water as the mobile phase has been reported by various authors.[13–15] The set-up is much less complicated than for the HPLC-FID coupling. The reason is that the column is usually connected with a long transfer capillary to the NMR, which fulfils different requirements. Firstly, the capillary acts as a linear restrictor, applying the necessary back pressure to keep the water in the liquid state. Secondly, the eluent need not be cooled down, because on its way to the NMR, ambient temperature is approached. Last but not least, the capillary separates the HPLC system from the NMR, because the main problem is that if metallic items are too near to the strong magnet in the NMR, they can damage the field and hence the magnet. Several authors used such an experimental set-up, the drawback being that a long transfer capillary adds to the extra-column band broadening.

The use of pure, or even buffered, deuterated water as the sole eluent is again very favourable for HPLC-NMR hyphenation, because binary mixtures with an organic co-solvent can interfere with the acquisition of NMR spectra by obscuring important regions of the spectrum. Although this problem can be greatly reduced by employing deuterated solvents, it cannot be eliminated. By using only deuterated water as the mobile phase, signals which would arise from protonated solvents like, for example, acetonitrile or methanol can be completely eliminated. This approach therefore provides ready access to that portion of the spectrum normally occupied by the solvent. Furthermore, deuterated water is much cheaper than deuterated acetonitrile.

In principle, the same rules apply for the elution of analytes which were already discussed in the previous section. This means that either isothermal or temperature-programmed separations can be performed. Again, the hydrophobicity of the stationary phase is very decisive in terms of eluting polar and non-polar sample constituents in one chromatographic run. Although most applications which were published some years ago employed polymeric stationary phases, the authors came to the conclusion that these phases generally required a higher temperature in order to achieve the same retention when compared to silica-based reversed-phase stationary phases. Unfortunately, at that time, the stability of silica-based phases was not as high as it is today. However, with the improved materials at hand today the situation is different. Furthermore, polymer-coated metal oxide stationary phases should be very well suited because they exhibit a low hydrophobicity, which means that the temperature need not be increased as much and hence, a possible on-column degradation of analytes is reduced.

A very impressive system, also including HPLC-NMR hyphenation, was constructed by Wilson and co-workers.[15] The authors coupled a high-temperature HPLC, using superheated water as the mobile phase, with a UV, an IR, a ^{1}H-NMR and an MS detector, extracting a maximum of information for the analysis of ecdysteroids in plant extracts. With the aid of this "hypernating" system, as it was termed somewhat tongue-in-cheek by the authors, they were able to characterize both pure standards and typical crude extracts of ecdysteroid-containing plants in on-flow mode, with full spectral characterization of the major ecdysteroids present. The sensitivity of the whole system was dictated by the IR and NMR spectrometers, rather than the UV and MS instruments, and were in the region of 100 µg for each compound on-column. The experimental layout of their system is depicted in Figure 8.3, and one of the examples showing the massive analytical information gathered from one chromatographic run is given in Figure 8.4.

Special Hyphenation Techniques

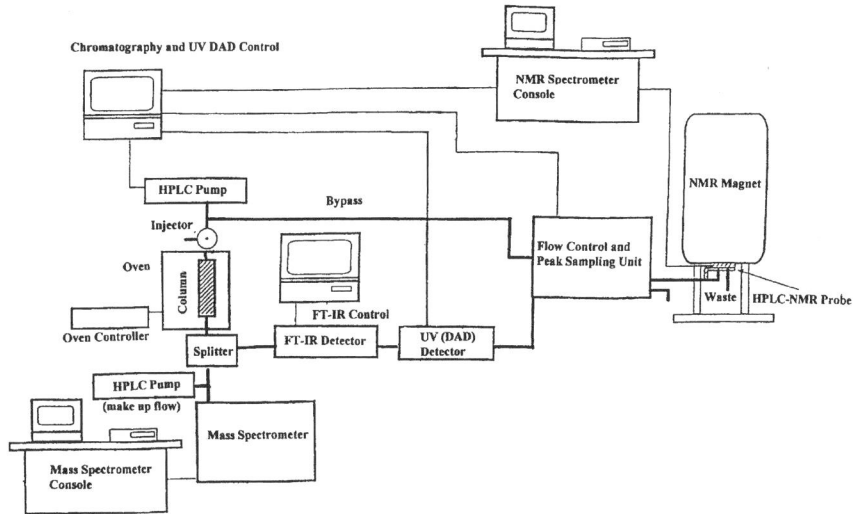

Figure 8.3 Experimental layout of the various spectrometers used in the superheated water HPLC-MS-IR-UV-NMR system.[15] (Reproduced with kind permission from ACS.)

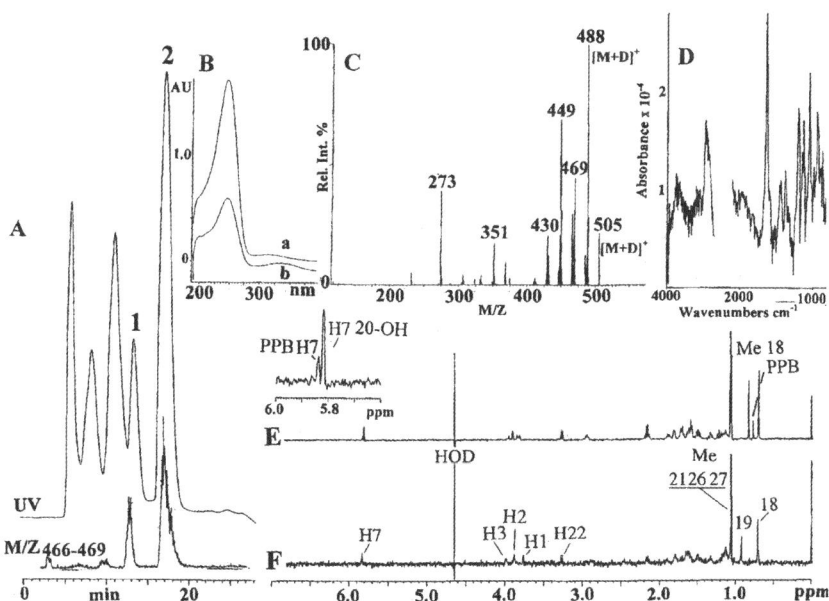

Figure 8.4 Chromatographic and spectroscopic data obtained for the extract of *S. nutans*.[15] (A) UV at 254 nm and mass (for m/z 466 to 469) chromatograms, peak identification: peak 1, integristerone A; peak 2, polypodine B and 20-hydroxyecdysone; (B) UV spectra for a, peak 2; and b, peak 1; (C) MS; (D) IR; (E) ^{1}H NMR spectra for peak 2 (including an expansion of the ^{1}H NMR spectrum for the C-7 protons); and (F) the ^{1}H NMR spectrum of peak 1. (Reproduced with kind permission from ACS.)

The authors clearly highlighted the difficulty of finding an optimal mobile phase for all detection systems, even if water is used as the sole eluent. While non-volatile buffers may well be suited to NMR, they are incompatible with MS detection. As was already mentioned, binary mixtures containing an organic co-solvent may be favourable for MS detection, but can interfere with the acquisition of NMR spectra.

Generally, the same conclusions about the applicability of superheated water chromatography with nuclear magnetic resonance spectroscopy can be drawn as for its hyphenation with flame ionization detection. The more complex the sample is, the more difficult it will be to achieve a baseline separation of all compounds in the mixture and to avoid a gradual wash-off of strongly hydrophobic compounds from the HPLC column. Therefore, sample preparation along with enrichment techniques will be of great help in reducing the complexity of the sample, and in increasing the amount of compound to be injected on column.

8.3 Isotope Ratio Mass Spectrometry

I guess that at the time of writing, this technique is relatively unknown in the chromatographic community, although it has an enormous potential. I will not go into detail about this technique, but will instead refer the reader to a forthcoming publication in the RSC series which is scheduled to appear in 2011.[16] In this book, the authors will describe the technology in depth and will also give some practical experiences which are far beyond the scope of this monograph. Nevertheless, I will focus on the most salient features, so that the reader will get an overview of this emerging technique.

Isotope ratio mass spectrometry (IRMS) is used to distinguish between compounds which are either faked or are synthesized by different pathways, but which are chemically identical except for their abundance ratio of stable isotopes. More specifically, isotope ratio mass spectrometry is used to measure very precisely the abundance of the heavy to the light isotopes of carbon. The rates at which heavier isotopes participate in chemical and physical processes are slightly different from those for lighter isotopes. The difference in rates leads to a subtle variation in the natural abundance of isotopes, owing to a variety of fractionation processes. The stable isotope composition of compounds is a function of their origin and history. Information about precise isotope ratios is very important in the nuclear, geological, agricultural, environmental and health sector. Up until now, gas chromatography has been mainly used in these studies to hyphenate the separation step with an isotope ratio mass spectrometer.[17,18]

Isotope ratios are expressed relative to reference standards, rather than being reported as absolute isotope values. The isotope community has established the δ-notation, which is the difference in the $^{13}C : ^{12}C$ isotope ratios of the sample and an internationally agreed standard normalized by the $^{13}C : ^{12}C$ isotope

Special Hyphenation Techniques 169

ratio of the standard. The resulting $\delta^{13}C$ value is given in ‰:

$$\delta^{13}C[‰] = [(R_{Sample}/R_{Reference}) - 1] \times 1000 \qquad (8.1)$$

Samples can be directly measured *versus* a reference gas, which is calibrated against the international reference.

However, the drawback of GC-IRMS analysis is that many analytes of interest cannot be measured without derivatization. This procedure is not only extremely time consuming, but also bears the risk of an isotope fractionation due to the derivatization process.[17] If it is not possible to correct the measured $\delta^{13}C$ values for this isotope fractionation, the results are useless. In a recent instrumental development to overcome this limitation, an LC interface for coupling high performance liquid chromatography to isotope ratio mass spectrometry was introduced.[19] In this system, all compounds are quantitatively converted into CO_2 while the analyte is still dissolved in the aqueous liquid phase. The chemical oxidation is typically performed by peroxodisulfate under acidic conditions. The CO_2 is removed from the eluent and entrained into a flow of helium by a miniature separation unit. This helium stream passes a water trap system and is then directed to the ion source of the IRMS *via* an open split assembly. For a better understanding, Figure 8.5 depicts the scheme of the so-called LC-IsoLink™.

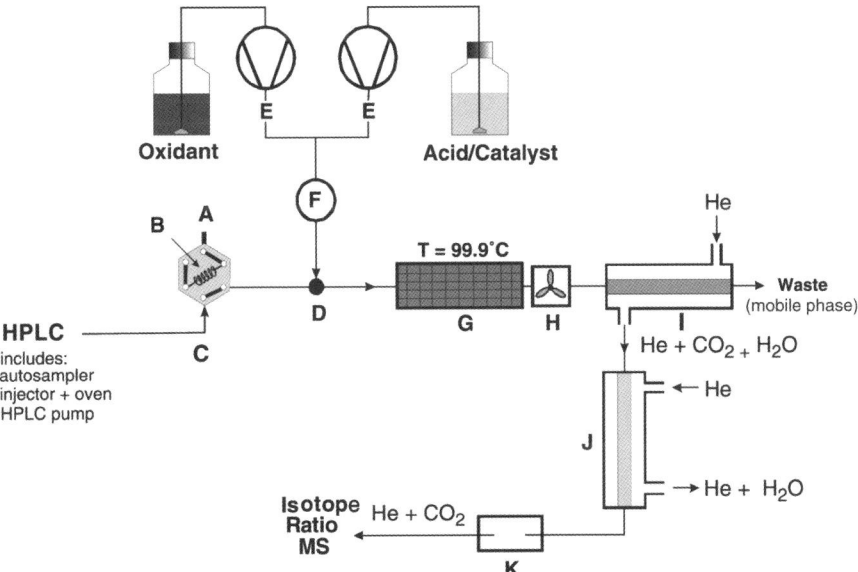

Figure 8.5 Scheme of the Finnigan™ LC IsoLink. A, Needle port; B, sample loop; C, 6-port valve; D, T-piece; E, two-head-pump; F, pulse damper; G, oxidation reactor; H, cooler; I, CO_2 separation unit; J, gas dryer; K, open split.[19] (Reproduced with kind permission from Elsevier.)

However, the hyphenation of liquid chromatography with isotope ratio mass spectrometry (IRMS) presents some restrictions which have to be overcome. First of all, the HPLC mobile phase must be totally free from organic or other carbon-containing compounds, because the full conversion to CO_2 is accomplished while the analyte is still dissolved in the mobile phase. Hence, only pure aqueous eluents can be used. The pH adjustment is possible by using inorganic buffers, such as phosphate buffers. All carbon-containing additives in the mobile phase must be avoided because they would yield a high background of the isotope ratio traces (m/z 44, 45 and 46). Again, the elution strength of water can only be changed by increasing the temperature, using the fact that water becomes less polar at higher temperatures, as has been discussed previously. Besides changing the temperature, the pH and the stationary phase can also be altered to achieve a different selectivity. One of the major drawbacks is that solvent gradients cannot be used. Here, the same restrictions apply as for hyphenation with the flame ionization detection.

I will now resume the discussion about the steroid application presented in Chapter 6.5.1. I would like to briefly repeat the conclusions which were made about the elution of these compounds at high eluent temperatures on a polybutadiene-coated zirconium dioxide stationary phase using a pure water mobile phase. It was possible to minimize the analysis time by increasing the flow rate up to $5\,ml\,min^{-1}$, so that all compounds eluted within 1.2 minutes (see Figure 6.9). However, this flow rate is too high for the introduction into the combustion oven of the IRMS system. Here, the flow rate has to be adjusted to around 300 to $500\,\mu l\,min^{-1}$. This is the first problem one has to overcome when non-polar compounds have to be eluted at a low flow rate and a high temperature. Therefore, a compromise has to be found between mobile phase flow rate and temperature. The higher the temperature, the higher the optimum linear flow rate required to operate the column at its van Deemter minimum. When analyzing polar compounds, elution can be achieved at ambient temperature and thus there is no need to worry about the optimum linear velocity. However, when analyzing non-polar compounds, such as steroids, under these conditions, it seems to be inevitable that the flow rate will still be in the B-term region of the van Deemter curve. There is an additional problem, however, which is directly related to high eluent temperatures. As I have shown in Chapter 5.1, column bleed can arise if the temperature is increased. This column bleed cannot be entirely suppressed, even if columns are used which are designed for high-temperature operation. Therefore, the elution of non-polar compounds requires a column which is stable at high eluent temperatures with low bleeding. Additionally, these columns should also have a low hydrophobicity when hydrophobic substances are analyzed. Otherwise, it may be impossible to elute strongly retained compounds from the column within an acceptable analysis time. Unfortunately, many columns which are used successfully in high-temperature liquid chromatography are very retentive. Therefore, a major objective of high-temperature liquid chromatography has been to find a suitable column which fulfils these requirements.

The work, which is currently carried out in the author's own laboratory, is also focused on this topic. For the elution of steroids, a polybutadiene-coated zirconium dioxide column gave the best results. We also compared a polystyrene–divinylbenzene stationary phase with the PBD-coated zirconium dioxide column in terms of bleed. We observed that the column bleed was constant for the latter column, while the PS–DVB column suffered from a poor run-to-run repeatability when operated in temperature-gradient mode. An additional advantage of the PBD-coated zirconium dioxide column was that it had a significantly lower hydrophobicity than polymeric columns based on polystyrene–divinylbenzene. This means that the temperature need not be increased as much as for the polymeric column to achieve the same retention.

It can be summarized that for the hyphenation of liquid chromatography with isotope ratio mass spectrometry, the column should have a low bleed during a temperature gradient as well as a low hydrophobicity if non-polar compounds are to be eluted. Otherwise, the temperature has to be increased to an extent where column bleed cannot be entirely suppressed.

We are confronted with another problem, however, in high-temperature LC hyphenated with isotope ratio mass spectrometry. What I have not yet discussed in detail is that steroids cannot be dissolved in pure water at ambient temperature, because their solubility is too low and most of the compounds will precipitate immediately at the desired concentration level. Therefore, an organic co-solvent has to be used to increase the solubility of the target compounds in the stock solution. This co-solvent can be methanol, acetonitrile or any other solvent which appears to be suitable. The problem is that carbon is introduced to the system. Usually, this solvent peak elutes at the void time and if UV detection is used, a problem will rarely occur. However, because every organic molecule which is introduced in the combustion oven will be converted to CO_2, a signal will be recorded. The background of this signal which is caused by the solvent peak may significantly interfere with the detection of the target compounds. Therefore, the same problem occurs in the hyphenation of high-temperature LC with IRMS as with flame ionization detection, in that the detector becomes "blind" against the mobile phase or the solvent peak, and there is no possibility of determining accurate delta values for the target analytes.

One solution is to use a second loop, which will be referred to as "solvent-cut", in order to separate the solvent peak from the analyte peaks. This technique helps to reduce the problem of interference of the solvent peak with that of the analytes, so that delta values can be precisely determined. Such a solvent-cut can be easily inserted in the standard hardware of the IRMS system.

In order to fully cut-off the solvent peak from that of the analyte peaks, the time between the elution of the solvent peak and the elution of the first analyte compound should be as high as possible. This means that the temperature programme should start at a "low" temperature, because this is the only option for maximizing the time between the elution of the solvent peak and the first compound and decreasing the overall analysis time of the late eluting compounds. Please note that it is simply not possible to adapt the methods I have shown in Chapter 6.5.1, because either the flow rate is too high, or the distance between the solvent peak and the first analyte peak is too low to allow an accurate determination of delta values. A baseline separation of all analytes by HPLC-IRMS was achieved by linearly increasing the temperature from 90 to 120 °C over 40 minutes. The resulting chromatogram is shown in Figure 8.6.

In order to highlight the difficulties which are encountered for HPLC-IRMS method development, I would like to compare the method I have presented above with a method which was recently published by Al-Khateeb and Smith.[20] In that paper, the authors present a separation of selected steroids on an XTerra MS C18 column. Using superheated water at 130 °C, 19-nortestosterone eluted at approximately 56 minutes on a 15 cm long column, while testosterone was not eluted from the column within a reasonable time. It has to be noted that the flow rate was adjusted to $3\,\text{ml}\,\text{min}^{-1}$, which is six times higher than the flow rate which can currently be employed for HPLC-IRMS hyphenation. When the column length was reduced three-fold, an elution of testosterone was only possible after 4 hours, at a flow rate of $1\,\text{ml}\,\text{min}^{-1}$ and with 5% methanol in the mobile phase. Hence, the results obtained by Al-Khateeb and Smith clearly highlight that the column has to be of a low hydrophobicity if an elution can be achieved within a reasonable time using water as the sole eluent. Otherwise, an elution of non-polar compounds is simply not possible with superheated water. Since 19-nortestosterone elutes before testosterone on the PBD-coated zirconia, as well as the hybrid C18 stationary phase, it can be expected that an elution of all the other compounds employed in our study would be impossible on a hybrid stationary phase using pure water, even if a very short column was used. The only alternative would be to apply extremely high temperatures up to the critical point of water in order to further decrease its polarity, but then a rapid degradation of the column may be expected.

Before I close this section, I would also like to add a few words regarding peak capacity. Nowadays, high peak capacities of approximately 900 can be obtained, even with one-dimensional separations, using long columns at elevated temperatures.[21] It can be easily deduced from observation of the chromatogram in Figure 8.6, that the peak capacity in our system is limited to approximately 10 to 20, even for a 15 cm long column. Since IRMS measurements are highly desirable for "real world" samples containing more than 20 compounds, as is the case of pesticide analysis, the question arises as to how peak capacity can be increased. In my opinion, there are principally two approaches which are feasible. The first is to use sample preparation techniques in order to fractionate the sample. These sub-samples then contain only a part of all the analytes included in the original sample. Also, the compounds in an individual sub-sample would have a similar polarity. Thus, an appropriate column with a suitable hydrophobicity could be chosen for each sub-sample. Another option would be to miniaturize the complete set-up using micro-bore or capillary columns, with an ID of 1.0 to 0.5 mm. As was shown experimentally by Molander *et al.*, capillary columns appear to be more

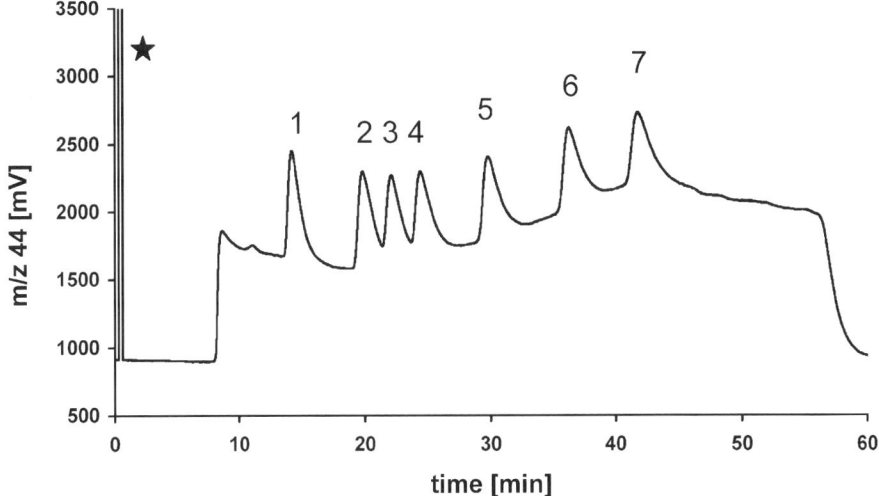

Figure 8.6 Temperature-gradient separation of selected steroids using IRMS detection. Temperature gradient: 90 to 120 °C in 40 min. Peaks: *CO_2 reference gas; 1, 19-nortestosterone; 2, testosterone; 3, epitestosterone; 4, *trans*-dehydroandrosterone; 5, etiocholan-3α-ol-17-on; 6, *cis*-androsterone; and 7, 5α-androstane-3α,17β-diol. Flow rate of mobile phase (water): 500 µl min^{-1}.

positively influenced by temperature gradients than larger bore columns with respect to chromatographic efficiency.[22] Furthermore, the authors showed that capillary columns possessed a higher robustness towards temperature programming. A concomitant advantage is that the optimal flow rate matched that required by the LC-IRMS interface, as was pointed out by Godin *et al*.[23] The authors stressed that for a 2.1 mm column, the optimum flow rate is approximately 145 µl min^{-1} at ambient temperature, but it reaches 600 and 1.500 µl min^{-1} at 100 and 200 °C, respectively. This means that if the temperature has to be adjusted to 200 °C to elute very hydrophobic compounds, additional band broadening occurs due to longitudinal diffusion if the flow rate is lower than the optimum linear velocity. However, decreasing the column's internal diameter also requires that the connecting tubing be shortened. Otherwise, the chromatographic resolution would be seriously affected. Moreover, if the column ID is reduced, column overloading must be avoided. However, it has to be considered that approximately 20 nmol of carbon are required for an accurate determination of the delta values in HPLC-IRMS. Hence, a further technical optimization of IRMS hardware seems inevitable.

I would like to end this section, with an example which illustrates how HPLC-IRMS is already a routine method. In 2008, Elflein and Raezke published a paper where they described an LC-IRMS method for the detection of honey adulteration.[24] A chromatogram for this method is given in Figure 8.7.

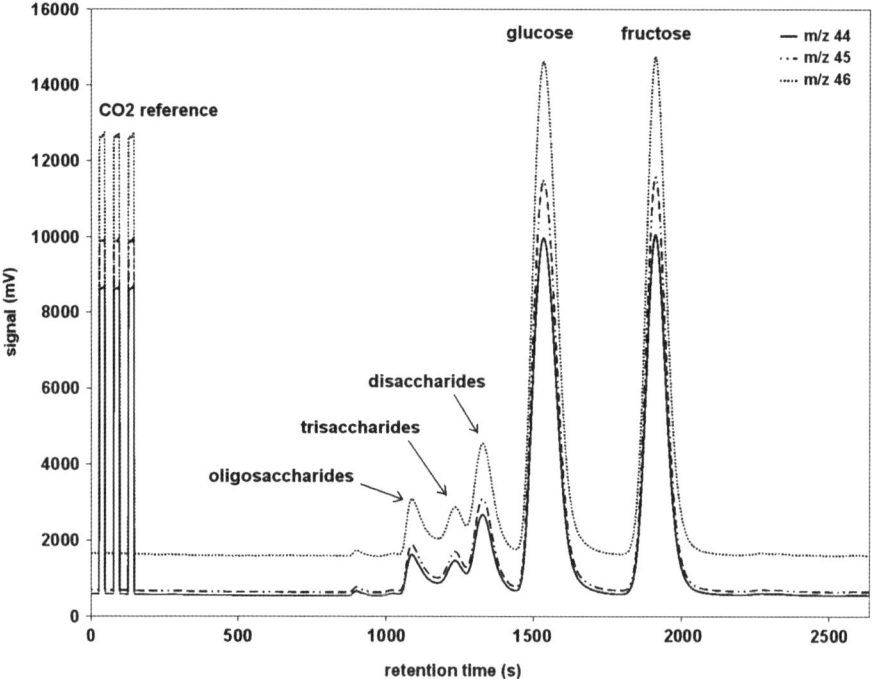

Figure 8.7 LC-IRMS chromatogram of adulterated polyfloral honey. $\delta^{13}C$ values: fructose –27.4 ‰, glucose –27.0 ‰, disaccharides –26.4 ‰, trisaccharides – 24.3 ‰, oligosaccharides –26.7 ‰. Adulteration: *ca.* 11 ‰ rice syrup.[24] (Reproduced with kind permission from EDP Sciences.)

Currently, this is the first and only worldwide accredited method for the detection of adulteration in honey. Please note that the separated compounds were highly polar and hence could be eluted at ambient temperature with a water only mobile phase. In contrast to this, the example I have given in Figure 8.6 unambiguously shows that even non-polar compounds, like steroids, can be eluted and separated on a suitable stationary phase without any derivatization. Hence, this technique is not only applicable for polar analytes like sugars, but also for non-polar sample constituents like steroids.

8.4 LC Taste®

A very interesting process, which has been patented by Symrise and also makes use of high-temperature HPLC, is the determination of gustatory active compounds in complex mixtures.[25] Flavour compositions contain at least two different sensorially active substances, such as synthetic, natural or nature-identical aromatic substances or plant extracts. Mostly, however, flavours are complex mixtures of many sensorially active components. Flavouring substances interact with the flavour receptors on the tongue and are responsible for

Special Hyphenation Techniques

the gustatory impressions. The proportions of the sensorially active substances in a flavour composition can vary enormously and they naturally have a strong influence on the overall sensory impression of the flavour composition. It is not the absolute amount of a sensorially active component in a flavour which is decisive, but its sensory contribution. Many sensorially important components in foodstuffs are not yet even known, since they are contained in only very small amounts, although they make a significant olfactory and/or gustatory contribution. More meaningful than the amount of a substance that is contained in a mixture are therefore the so-called "odour activity values" or "taste activity values", which are defined as the quotient of the concentration of a sensorially active component and its olfactory or gustatory threshold value.

What does such a system set-up for this process look like? A schematic illustration is given in Figure 8.8.

The HPLC system consists of two pumps for solvent gradient elution, a column oven with a suitable stationary phase and the option to cool down the mobile phase after the separation step. Afterwards, the mobile phase is introduced into a "technical" detector which can be a refractive index, an evaporative light-scattering, a UV, a DAD or a mass spectrometric detector. Additionally, this set-up contains a flow-splitter so that the eluate can be directly tasted by a person. The only problem is that in contrast to "conventional" HPLC, the mobile phase must not contain harmful or toxic substances. Usually, acetonitrile, methanol or tetrahydrofuran are used as organic co-solvents for solvent gradient elution. However, all of these solvents have severe side effects if swallowed by a human being. Therefore, these solvents must be removed from an eluted fraction before a sensory evaluation. Such a removal requires the use of mostly thermal or extractive processes such as distillation or freeze-drying. However, this may result in a change in the composition of the flavouring substances contained in the eluted fraction due to thermal and or oxidative influences. The isolated sensorially active compounds must then be taken up in a harmless solvent in order to be sensorially evaluated by means of dilution analysis. The dilution analysis and the subsequent sensory evaluation are not automated processes and hence, the complete procedure is extremely time-consuming. Therefore, with the set-up depicted in Figure 8.8, it is now possible to perform an on-line sensory evaluation of flavour compositions with a human sensor and conventional HPLC detectors during one chromatographic run.

As becomes evident, pure water is the preferred eluent because it is not toxic and will not interfere with the detection process of the human being or the technical detector. However, as was also pointed out before, the elution strength of pure water at elevated temperatures is not sufficient for the elution of non-polar components. Therefore, the addition of modifiers which increase the elution strength of the mobile phase are necessary if these compounds are to be analyzed. It is clear that typical organic solvents are strictly forbidden if the eluate is to be directly tasted by a human being. Nevertheless, it can be shown that ethanol is a very convenient co-solvent which – up to a certain concentration – has no negative impact on sensory impression and also

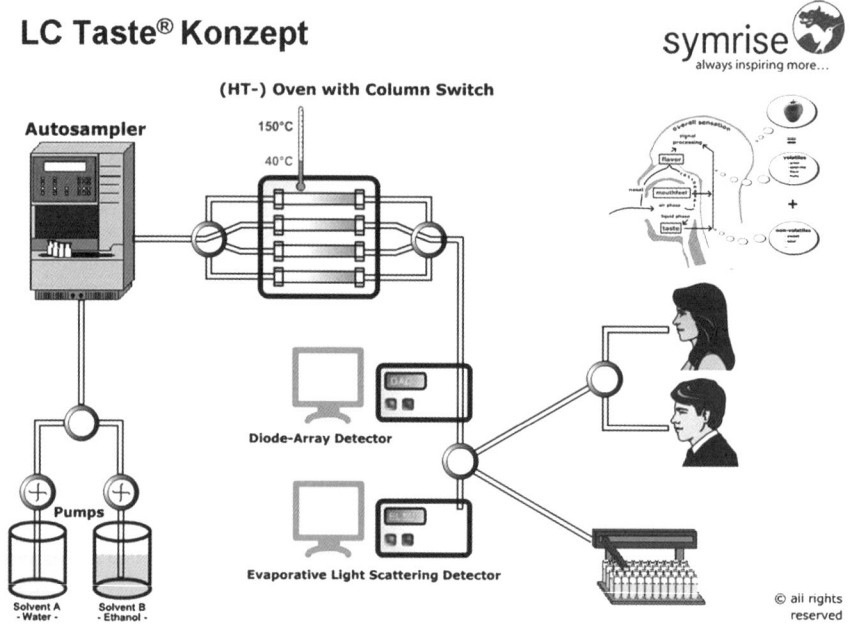

Figure 8.8 Concept of LC taste[R]. (See the text for further details.) (Reproduced with kind permission from Symrise.)

significantly enhances the elution strength of the mobile phase if solvent gradient elution is applied.

In contrast to the hyphenation techniques I have described before and which rely only on pure water as the sole eluent, the concept of LC taste[R] allows for the combined use of temperature and solvent programming. This is a decisive advantage because the elution strength can be increased by a simultaneous change in temperature and in mobile phase composition. Moreover, there is a far greater chance to elute all compounds in the mixture within one chromatographic run. In addition, a band compression due to solvent-gradient programming is possible as was already outlined in Chapter 6 and hence, the peak capacity can be tremendously increased. However, the ethanol concentration in the mobile phase should preferably be in the range of 5 to 30 wt%. Therefore, a concomitant temperature gradient is necessary so that the concentration of ethanol will be as low as possible. The procedure of how to optimize both types of gradients has been laid out in detail in Chapter 6.5.3. This application highlights that, besides optimizing selectivity, temperature programming can be used in combination with solvent programming to exhibit the unique feature of reducing organic co-solvent to a minimum.

Another requirement for LC taste[R] is that the temperature of the mobile phase leaving the column should be no higher than 40 °C. Otherwise, the burning recorded by the human operator would not be attributed to the flavour of the separated compound but to the high eluent temperature. Therefore, a heating system with a module for eluent cooling as described in Chapter 3 is mandatory.

8.5 Drug Screening

The screening of synthetic and natural chemical sources is usually the starting point in drug discovery. High-throughput screening technologies have been developed and implemented that are able to test ten thousands of compounds or more per day for their activity in various assay types, ranging from receptor-binding and enzyme-inhibition to whole cell assays. While such high-throughput techniques are highly efficient for the screening of pure compound samples (libraries), the screening of complex mixtures is more demanding. A different approach based on high-temperature HPLC and focused on the determination of biologically active compounds in complex mixtures has been presented by de Boer and Irth.[26]

Examples of complex mixtures in drug discovery include samples originating from natural products, reaction mixtures from solution-phase combinatorial chemistry, and *in vitro* or *in vivo* metabolic profiling. In all cases, non-active sample constituents, at widely differing concentration ranges, are present next to an unknown number of pharmacologically active compounds. The main difficulties encountered are the correlation of biological activity with chemical analysis data for the rapid identification of active compounds and the presence of matrix components that interfere with the assay readout. Identification requires fractionation, mostly performed by off-line liquid chromatography, in combination with fraction collection. The whole process of screening and fractionation must be repeated until the bioactive compound against the molecular target is isolated. It is obvious that this process can be very laborious, time consuming and error prone.

In order to overcome these drawbacks, an on-line coupling of a separation step with the biochemical assay would be very advantageous. It was with this in mind that de Boer and Irth developed a method in which an HPLC separation was hyphenated to an enzyme assay. The biochemical assay was based on a continuous-flow enzyme-substrate reaction and the subsequent detection of reaction products by electrospray-ionization mass spectrometry (ESI-MS). Inhibition of enzyme activity by compounds eluting from the HPLC column resulted in a temporary change in product concentration, which was determined by ESI-MS as a negative peak in the extracted ion chromatograms (EIC) of the products.

The compounds which eluted from the column were mixed with an enzyme in order to monitor if there was a specific inhibition of the enzyme. Cathepsin B was used as a model enzyme which catalyzes the hydrolysis of a substrate called Z-FR-AMC, yielding two products, Z-FR and AMC. The enzyme assay was performed in a continuous-flow, post-column reaction detection system comprising two reaction coils. In reaction coil A, analytes eluting from the HPLC column interacted with cathepsin B in the absence of the substrate. The concentration of active cathepsin B was monitored by the addition of the substrate (Z-FR-AMC) in reaction coil B. Inhibition of cathepsin B was detected by monitoring changes in the concentration of the reaction products AMC and Z-FR using ESI-MS. For a better understanding, this system is given in Figure 8.9.

However, a limitation of this approach is that the enzyme assay can only be coupled on-line if the concentration of the organic modifier in the mobile phase does not exceed a threshold where the enzymatic activity will be lost. Again, this is the reason why high-temperature HPLC was used. Increasing the temperature of the mobile phase offered the possibility of reducing the amount of the organic co-solvent, so that the enzymatic activity will be maintained after the separation step. This drawback could also be circumvented by the post-column addition of an aqueous solution, but this has the disadvantage of diluting the effluent and requiring a more complex analytical system.

The authors showed that biochemical assays could be performed in the presence of organic modifier concentrations up to 15%, as long as the reaction time did not exceed 3 to 5 minutes. Similar results were obtained for cathepsin B, in which the presence of 10% methanol in the enzyme-substrate reaction – which corresponded to 20% methanol in the column – led to an 11% decrease in product formation.

It should be noted that the use of 20% methanol in the mobile phase can be regarded as very high, if temperature programming is used in combination with solvent programming. Again, the concomitant use of these two gradient

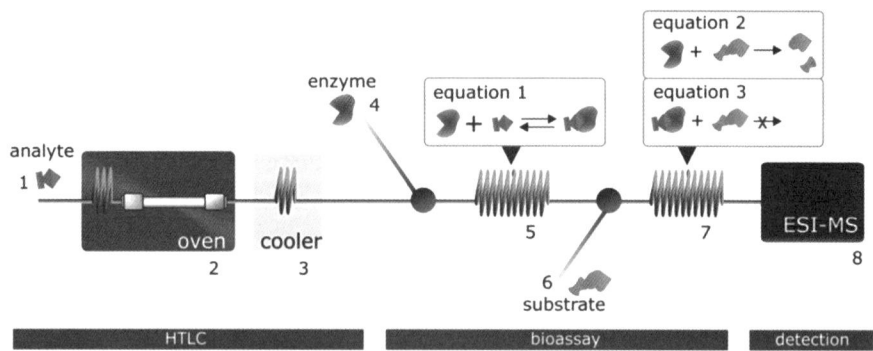

Figure 8.9 Overview of the continuous-flow system. 1, sample introduction; 2, HTLC oven containing the heating coil and column; 3, cooler; 4, superloop with enzyme solution; 5, bioreactor A; 6, superloop with substrate solution; 7, bioreactor B; and 8, ESI-MS. The enzyme continuously converts the substrate into products (eq. 2) if no bioactive compounds are eluting from the column. Bioactive compounds present in the effluent bind to the enzyme (eq. 1) resulting in a decrease in product turnover (eq. 3). The reaction products are continuously detected by ESI-MS.[26] (Reproduced with kind permission from ACS.)

techniques allows for a band compression, leading to higher peak capacities than if pure water at elevated temperatures was used. Furthermore, the elution strength is much higher than for a pure water mobile phase. So the same conclusions can be drawn as for the concept of LC taste® which was described in the previous section. A decisive advantage of the technique described here is that the organic co-solvent is not limited to ethanol. In principle, all of the solvents that I have listed in Table 1.1 can be used. As was also noted earlier, complex mixtures usually contain different compounds with a wide range of polarities. Therefore, the right strategy is to find a mobile phase which, at a given maximum temperature, exhibits the lowest polarity. This maximizes the possibility that all sample components will be eluted from the column. Once again it should be reiterated that the use of an appropriate HPLC column, which exhibits both good selectivity and low retentivity, is very important. In addition, the data which is collected in Appendix C may be very useful in determining the polarity of different solvent systems and their dependence on temperature.

The authors also studied the influence of temperature on the separation of cathepsin B inhibitors and flavonoids. They noted that comparable retention factors were obtained if the separation was carried out at 25 °C using 80% methanol, and at 185 °C using only 10% methanol. A PLRP-1 column consisting of polystyrene–divinylbenzene was used for these experiments. As was pointed out earlier and is also mentioned by the authors, this column exhibited a high hydrophobicity. Hence, the addition of 10% methanol was necessary to achieve an elution of sample analytes within a reasonable time.

Presumably, this separation could have been carried out on a PBD-coated zirconium dioxide stationary phase with pure water, but this has yet to be shown. Although the peaks eluted with an acceptable symmetry, they appear relatively broad. In addition to the fact that the column was not operated at the van Deemter minimum, polymeric columns suffer from a low efficiency, as has also been pointed out in Chapter 5. Moreover, a band compression may have been obtained if solvent gradient programming had been applied. However, it is quite remarkable that an eight-fold reduction in methanol was possible by increasing the temperature from 25 °C to 185 °C.

In contrast to methanol, the higher chain alcohols should exhibit a lower polarity and thus a higher elution strength at the same temperature and modifier concentration, as has been demonstrated in Chapter 4.3 (see Figure 4.16). It would be worthwhile studying the effects of alternative solvents in detail, in order to find the solvent system which demonstrates the best elution characteristics for the separation of the mixture, without compromising enzyme activity or detection sensitivity for ESI-MS. It should also be noted that the eluent needs to be cooled down after the separation step. Otherwise, the enzyme activity will be decreased, because the assay works best at around ambient temperature. Additionally, it should be investigated if a constant response is obtained if a solvent gradient is run from 5 to 15% of the organic co-solvent. It may be that the enzyme activity will run through a maximum, and after the optimal modifier concentration is exceeded the enzyme activity will decrease.

Current activities in the author's own laboratory are focused on establishing this technique for the screening of the toxic and allergenic potential of house dust samples. From the discussion in this chapter, it is evident that a lot of basic research is needed in order to find the optimum conditions for this approach. However, once the technical difficulties are solved, this technique could be a real alternative to time-consuming, off-line procedures.

References

1. T. Teutenberg, S. Wiese, P. Wagner and J. Gmehling, *J. Chromatogr., A*, 2009, **1216**, 8480.
2. T. Teutenberg, J. Tuerk, M. Holzhauser and S. Giegold, *J. Sep. Sci.*, 2007, **30**, 1101.
3. T. Teutenberg, K. Hollebekkers, S. Wiese and A. Boergers, *J. Sep. Sci.*, 2009, **32**, 1262.
4. D. J. Miller and S. B. Hawthorne, *Anal. Chem.*, 1997, **69**, 623.
5. E. W. J. Hooijschuur, C. E. Kientz and U. A. T. Brinkman, *J. High Resolut. Chromatogr.*, 2000, **23**, 309.
6. Y. Yang, A. D. Jones, J. A. Mathis and M. A. Francis, *J. Chromatogr., A*, 2002, **942**, 231.
7. T. Yarita, R. Nakajima, S. Otsuka, T. A. Ihara, A. Takatsu and M. Shibukawa, *J. Chromatogr., A*, 2002, **976**, 387.
8. R. Nakajima, T. Yarita and M. Shibukawa, *Bunseki Kagaku*, 2003, **52**, 305.
9. D. Guillarme, S. Heinisch, J. Y. Gauvrit, P. Lanteri and J. L. Rocca, *J. Chromatogr., A*, 2005, **1078**, 22.
10. Y. Yang, T. Kondo and T. J. Kennedy, *J. Chromatogr. Sci.*, 2005, **43**, 518.
11. N. Wu, Q. Tang, J. A. Lippert and M. L. Lee, *J. Microcolumn Sep.*, 2001, **13**, 41.
12. B. A. Ingelse, H.-G Janssen and C. A. Cramers, *J. High Resolut. Chromatogr.*, 1998, **21**, 613.
13. R. M. Smith, O. Chienthavorn, I. D. Wilson, B. Wright and S. D. Taylor, *Anal. Chem.*, 1999, **71**, 4493.
14. D. Louden, A. Handley, S. Taylor, I. Sinclair, E. Lenz and I. D. Wilson, *Analyst*, 2001, **126**, 1625.
15. D. Louden, A. Handley, R. Lafont, S. Taylor, I. Sinclair, E. Lenz, T. Orton and I. D. Wilson, *Anal. Chem.*, 2002, **74**, 288.
16. T. C. Schmidt and M. A. Jochmann, *Compound Specific Isotopic Analysis*, RSC Publishing, Cambridge, 2011, in preparation.
17. V. Ferchaud, B. Le Bizec, F. Monteau and F. Andre, *Analyst*, 1998, **123**, 2617.
18. P. M. Mason, S. E. Hall, I. Gilmour, E. Houghton, C. Pillinger and M. A. Seymour, *Analyst*, 1998, **123**, 2405.
19. M. Krummen, A. W. Hilkert, D. Juchelka, A. Duhr, H. J. Schluter and R. Pesch, *Rapid Commun. Mass Spectrom.*, 2004, **18**, 2260.
20. L. A. Al-Khateeb and R. M. Smith, *Anal. Bioanal. Chem.*, 2009, **394**, 1255.

21. P. Sandra and G. Vanhoenacker, *J. Sep. Sci.*, 2007, **30**, 241.
22. P. Molander, R. Olsen, E. Lundanes and T. Greibrokk, *Analyst*, 2003, **128**, 1341.
23. J. P. Godin, G. Hopfgartner and L. Fay, *Anal. Chem.*, 2008, **80**, 7144.
24. L. Elflein and K. P. Raezke, *Apidologie*, 2008, **39**, 574.
25. M. Roloff, H. Erfurt, G. Kindel, C.-O. Schmidt and G. Krammer, *Process for the Separation and Sensory Evaluation of Flavours*, Symrise GmbH & Co. KG, Int.-Publ.-No.: WO 2006/111476 A1, World Intellectual Property Organization, 2006.
26. A. R. de Boer, J. M. Alcaide-Hidalgo, J. G. Krabbe, J. Kolkman, C. N. van Emde Boas, W. M. Niessen, H. Lingeman and H. Irth, *Anal. Chem.*, 2005, **77**, 7894.

CHAPTER 9
Critical Outlook and Future Prospects

Although guesses or predictions concerning the outlook of a new technique or the extent to which a method will evolve in the future are frequently unsuccessful, I will, nevertheless, attempt this "dangerous" undertaking.

I hope that I have been able to demonstrate how the use of elevated or high temperatures in liquid chromatography could be a boost not only for everyday analyses, but also for the new hyphenation techniques which were presented in the previous chapter. The instrumentation, which is necessary for an efficient use of high-temperature HPLC, is now commercially available. Instrument manufacturers will continue to improve their products, however, and to add new features to their existing systems. What I have also noted is that although the HPLC hardware is quite advanced, column technology is lagging behind. The optimum stationary phase for high-temperature liquid chromatography has yet to be developed.

9.1 Pellicular Particles

There is a number of different approaches which may prove useful in the future. It is quite helpful to screen the "old" literature and to carefully examine the work of the pioneers in chromatography. Unfortunately, in today's world, there is really an information overload. Important information from the past is often simply overlooked, because it is assumed that papers which were written more than twenty years ago will not have a big impact on the problems we are facing today. For the majority of all publications, this might be true, especially since the technical advances in analytical sciences have gained a momentum in the last few years and new technologies have been introduced to the market. But when I searched the literature I came across a technology, which was introduced more than 20 years ago but was not accepted by the market. In 1993, Chen and Horváth reported an experiment where they compared pellicular particles which had a solid core with the respective fully porous particles.[1] They found out that columns packed with 2 μm pellicular ODS silica showed no degradation after 1000 hours of operation at 120 °C. Columns packed with highly cross-linked, non-porous polystyrene beads could withstand high temperatures up to 200 °C for at least 300 hours. The integrity of a column packed with commercial porous C-18 silica beads, however, was impaired even after a short exposure to hydro-organic eluents at 120 °C. In contrast to this, columns packed with PLRP-S polystyrene–divinylbenzene beads (5 μm, 300 Å) showed only a slight loss in efficiency after 200 hours of operation at 120 °C.

It seems that nobody really cared about these results and even people working in the field of high-temperature HPLC concentrated on other materials, such as metal oxides or polymeric packings. The problem was that the loading capacity of pellicular packings was said to be low when compared to fully porous particles, because the available surface area is lower than that of fully porous particles. Since the pharmaceutical industry is faced with the problem of detecting impurities and degradation products of the active pharmaceutical ingredient (API) at very low concentrations, a high loading capacity of a column is a prerequisite. Therefore, pellicular packings never really entered the stage of the life science world.[i] However, nowadays the concept of pellicular packings has been revisited. They are now called fused-core packings, but in effect the term pellicular is more appropriate. We have recently reported the temperature and pH stability of some commercial stationary phases, including a silica-based ODS stationary phase, which were very rapidly degraded.[2] However, colleagues from BayerCrop Science also used this material, but instead of the manufacturer's porous packing they employed the corresponding pellicular particles. We were astonished when they reported that the material could be considered very stable at high eluent temperatures.[3] As the methods used for column testing did not allow for a fair comparison, the experiments need to be repeated under the test conditions I have described in Chapter 5. Please remember that hybrid particles based on silica gel are very rugged against thermal stress. If it can be confirmed that pellicular particles exhibit an even higher stability than the fully porous particles, a combination of the two techniques would be even better. Of course, the problem of a lower loadability remains, but the aim is to improve column lifetime for high-temperature HPLC. Undoubtedly, pellicular particles exhibit less back pressure and separations can be carried out much faster because the diffusion pathway is significantly reduced. These are all very positive features, so that a broader acceptance of pellicular particles could also be a boost for high-temperature liquid chromatography.

---

[i] Please note that for the analysis of sample mixtures in which the range of component concentrations is wide, *e.g.* exceeding $1-1 \times 10^8$, it is important that the column has both an excellent efficiency and a large loading capacity. To achieve the good resolution needed for the detection and quantification of trace components, the column should not be overloaded with the major components, which then would give asymmetrical peaks due to non-linear thermodynamic effects.

9.2 Capillary and Nano HPLC

The next topic I would like to have a closer look at is miniaturization. Today we are under pressure to significantly improve the overall cost effectiveness in the laboratory. However, a crisis can also have some positive effects. As the cost of resources steadily increases, we are forced to implement new technologies. Miniaturization could be very effective in fulfilling both these requirements. As has already been mentioned in the text, mass spectrometry is becoming ever more important in all fields of the life sciences. For a long time it was not easy to overcome the hurdles and limitations of LC-MS, in contrast to GC-MS which was well established and had the advantage that the mobile phase did not present any problems for the MS. A mass spectrometer is not designed to "swallow" large amounts of mobile phase and then detect only a few molecules in the gas phase. The problem has always been how to get rid of the mobile phase. Instrument manufacturers now offer a variety of MS systems which are compatible with relatively high flow rates. But what is meant by "relatively high flow rates?" The most widely used ionization process today is electrospray ionization, because it has a high efficiency for both large and small molecules. Therefore, it would be beneficial to reduce the flow rate in order to increase the detection sensitivity. In principle, capillary or nano-HPLC is the ideal technology for combining both a low consumption of resources, including energy and waste, with a high detection efficiency due to the low amount of mobile phase which is introduced into the mass spectrometer.[4,5] First of all, the solvent consumption is minimized in a capillary system when the pumps deliver the mobile phase in the $\mu l \, min^{-1}$ range. Furthermore, if there is only a limited amount of sample material, the injection of a volume as small as 50 nanoliters is technically feasible.

A paper by Granger and Wilson in 2005 can serve as a good reference because the authors compared a conventional HPLC-TOF-MS and a capillary-TOF-MS hyphenation.[6] In principle, the overall pattern of peak distribution provided by both techniques was not dissimilar. However, the capillary method revealed the presence of many more components than the conventional separation. In addition, even though much less material was injected onto the column in the capillary analysis, there was a distinct improvement in sensitivity. Whilst both methods provided similar retention times, the capillary technique was approximately 100 times more sensitive. The reduction in the numbers of solvent ions entering the MS instrument when using capillary HPLC also resulted in a significant increase in the quality of the data.

The hyphenation of capillary chromatography with mass spectrometry can be considered the perfect marriage of two techniques, since the low flow rates are highly compatible with unassisted electrospray ionization (ESI). This in turn leads to a higher sensitivity which improves detection for a range of compounds. However, there are also drawbacks which are extensively discussed and which have prevented capillary liquid chromatography from making a real breakthrough in the pharmaceutical industry. Often, problems are observed in terms of the retention time reproducibility, because many systems are not able to deliver a constant flow. In many cases, a flow splitter is used to reduce the flow rate, which means that the overall solvent consumption equals conventional HPLC systems. Also, solvent-gradient reproducibility is considered a major problem, because the low flow and, thus, the solvent mixing cannot be controlled precisely. Nevertheless, new systems especially designed for capillary or nano liquid chromatography have entered the market and are widely used in the biotechnology industry.[7,8]

The use of nano columns with an internal diameter down to 75 µm requires a system which is fully optimized in terms of dead volume, because this has a detrimental effect on extra-column band broadening and the overall efficiency of the separation and detection. The fact that normal-bore columns, with a diameter between 3 and 4.6 mm, are much easier to handle than capillary columns has also led to a widespread reluctance to use capillary liquid chromatography instead of conventional HPLC.

Apart from this, there are distinct advantages of using capillary or even nano columns, which were already mentioned in Chapter 3. Operating a column at a very high pressure leads to frictional heating and thus radial temperature gradients. By reducing the inner diameter of the column, these effects will become negligible. In addition, capillary columns should be much better suited for temperature programming than normal-bore columns, which is why many studies on temperature programming have focused on capillary columns.[9–17] Although I have clearly demonstrated that contact heating is very efficient, even for a fast and precise heating of columns with an internal diameter of 4.6 mm, much less thermal mass needs to be heated when using capillary columns. In this respect, the miniaturization of the column is very favourable.

Despite the technical implications, the majority of the techniques I have described in Chapter 8 would also benefit from capillary HPLC. The flame ionization detector is best operated if a low flow rate is used. This is typically around 200 µl min^{-1} to guarantee an efficient droplet formation and to prevent the flame from extinguishing. The hyphenation of HPLC with a biochemical assay and an isotope ratio mass spectrometer would also benefit from a reduction in the column diameter.[ii] Only the concept of LC taste® requires that there be a large amount of sample which can be tasted by a human being, and thus capillary chromatography makes no sense for this technique. Certainly, if flow rates in the lower µl min^{-1} range are applied, the low amount of each component in a chromatographic peak would be far too small to trigger a signal. Here, a scale-up to semi-preparative HPLC is more advantageous. Nevertheless, for all other hyphenation techniques the reduction of the column ID would be favourable when operating the column at high temperatures. The reason is that the optimum linear velocity increases at high eluent temperatures, which might be too high for the detector. By decreasing the column ID, the optimum flow rate at a constant temperature is lower. As a result, capillary columns can be used at high temperature and low flow rates which are

---

[ii] For isotope ratio mass spectrometry, some technical difficulties prevail which currently prevent the use of capillary columns.

compatible with the detection system without sacrificing chromatographic efficiency. This is most important when solvent programming cannot be used to achieve a band compression.

Although I have now discussed why it is beneficial to use capillary liquid chromatography in terms of the detection sensitivity, many practitioners remain sceptical about the robustness of these systems. When the literature is screened, it is striking that the peak widths which are produced in capillary liquid chromatography are much wider than the peaks from an ultra-high pressure system used in combination with small-particle packed columns of between 1 and 2 mm ID. Often, peak widths are around 5 to 30 seconds when capillary columns are used, while the peak width can be easily decreased to a second when advanced ultra-high pressure systems operated at high temperature and high flow rates are employed. Furthermore, it is critical to consider the time the separated eluent bands take to be transported through the connection tubing, from the point at which the eluent leaves the column up to the point at which it is introduced into the detector. In many cases, a hyphenated HPLC system is not optimized to minimize the dead volume. Even many instrument manufacturers seem to pay little attention to this issue. If we look to some standard configurations of how an LC system is coupled to a mass spectrometer, then it is striking that there is often a considerable length of connecting tubing between the outlet of the column and the inlet of the mass spectrometer. However, when 2 mm columns are used and the flow rate is adjusted to 500 μl min^{-1}, good chromatograms with peak widths of about 1 to 5 seconds can be obtained. If a capillary or even nano-HPLC system were be used, the results would be worse. The problem is that for a separated eluent band it takes a lot more time to travel through the transfer tubing between the column and the mass spectrometer when the flow rate is low. Depending on the heating system which is used, the length of this transfer capillary could be even longer if the heating system cannot be placed to provide an optimal connection of the column with the detector. Hence, the use of heating ovens specially designed for capillary or nano-columns in combination with MS detection appears to be necessary. One solution to this problem could be the use of so-called mini-column ovens which can be positioned directly in front of the inlet of the mass spectrometer. Such a set-up is shown in Figure 9.1 and was employed for the hyphenation of a capillary HPLC system with a time-of-flight mass spectrometer in the author's own laboratory.

The experiments were carried out using a capillary system from Eksigent based on microfluidic flow control. This system enabled the use of flow rates between 1 and 30 μl min^{-1} without flow splitting, and was optimized for 300 μm ID columns. Although the system was equipped with an integrated column thermostat, the connection of the column with the inlet source of the mass spectrometer was achieved by using an external heating device. Here, an HPLC Mini Column oven from AμMass, especially designed for LC-MS systems was used. The column was clamped between two aluminium blocks so that an efficient heat transfer was guaranteed. In effect, the Mini column oven is based on the same principle as the block-heating oven described in Chapter 3.

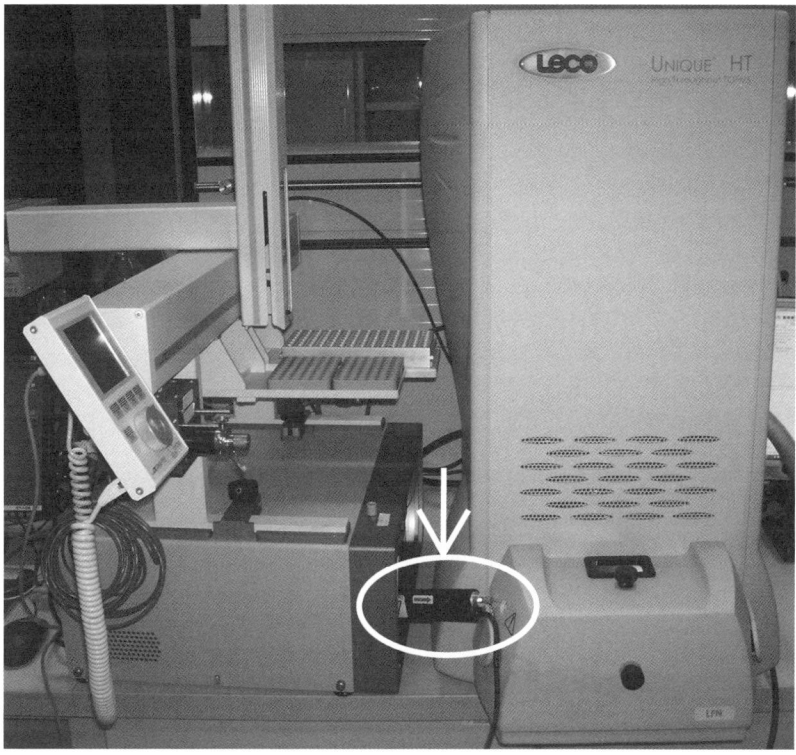

Figure 9.1 Hyphenation of a capillary HPLC system with a time-of-flight mass spectrometer using a mini-column oven.

The assembly of the column with the injection port was accomplished by using a PEEKSIL capillary (11 cm; 360 μm OD; 30 μm ID). The column was connected to the source of a LECO Unique HT TOFMS mass spectrometer, which was equipped with a low nebulizer source by using a New Objective standard coated Taper Tip (20 cm; 360 μm OD; 20 μm ID). In order to test this set-up, a standard mixture of 400 pesticides at a concentration of 100 pg μl^{-1} of each analyte was injected. Figure 9.2 shows a typical chromatogram which was obtained when using the system set-up as described above.

The total ion chromatogram (TIC) and the extracted ion chromatograms are shown. When the data was processed automatically, the software was able to find the majority of the compounds present in the sample. In Figure 9.3, a detail of the chromatogram of Figure 9.2 is shown. Within the time interval between 9 min 20 s and 10 min 40 s, approximately 50 compounds were found. Every analyte which was automatically detected by the software is highlighted by a peak marker. The peak width of each compound was about 6 to 8 seconds, so there is room for further improvement. Using a scan speed of eight spectra per second, an average of 80 data points across a peak was obtained which allows for accurate quantification.

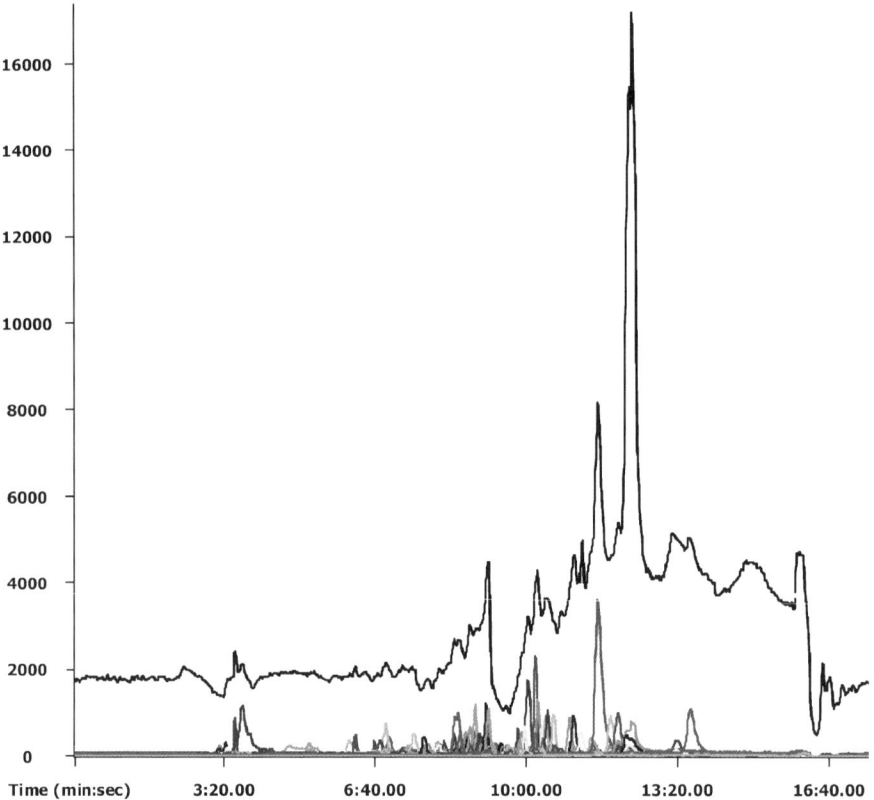

Figure 9.2 Chromatogram of a pesticide mixture. The upper curve represents the total ion chromatogram, below are the extracted ion chromatograms. Chromatographic conditions: column: Phenomenex Synergi Fusion-RP (100 mm × 0.3 mm ID; 100 Å, 2.5 µm); solvent A: water (containing 0.1% formic acid); solvent B: methanol (containing 0.1% formic acid); solvent gradient: 30% solvent B for 1 min, then 30 to 100% solvent B in 9 min, and hold at 100% solvent B for 5 min; flow rate: 3 µl min^{-1}; temp.: 40 °C. Detection: instrument: LECO Unique HT TOFMS; ESI voltage: 3.2 kV (positive); mass range: m/z 20 to 1000; scan speed: 8 spectra s^{-1}.

Compared with classical approaches by triple-quadrupole MS, time-of-flight (TOF) instruments allow for the screening of a high number of known and unknown compounds within one chromatographic run.[iii] This is also confirmed by other authors who have used the hyphenation of capillary LC-ESI-TOF-MS. As an example, Andersen and co-workers developed a method to detect and characterize high-molecular-mass hindered amine light stabilizers.[18]

[iii] However, several providers have recently commercialized new quadrupole-based instruments with improved full-scan acquisition rate (up to 10 000 m/z per second), reduced dwell time as low as 5 ms for selected-ion monitoring mode (SIM) and rapid (ca 20 ms) polarity, as well as ionization-mode switching (for simultaneous ESI+/ESI−, APCI+/APCI−, ESI/APCI and APCI/APPI operations).

Critical Outlook and Future Prospects 189

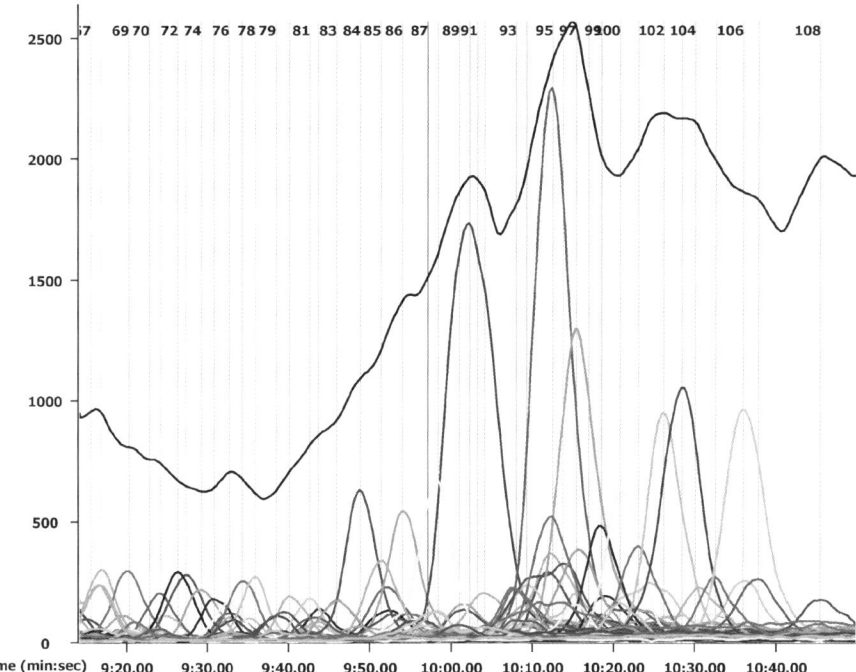

Figure 9.3 Chromatogram of a pesticide mixture. Every analyte which has been automatically detected by the software is highlighted by a peak marker. The upper curve represents the total ion chromatogram, below are the extracted ion chromatograms. (For other details: see Figure 9.2.)

They also used an approach similar to that described above and concluded that the hyphenation had many advantages which were due to the low flow rate in the capillary column. They employed temperature programming and highlighted that this approach may provide an alternative, or even superior technique, to the more traditional approach of gel-permeation chromatography, due to the improved chromatographic resolution which could be achieved.

9.3 Comprehensive Two-Dimensional Liquid Chromatography

An important question remains concerning the enhancement of chromatographic resolution and peak capacity. Nowadays, two-dimensional liquid chromatography is being extensively discussed as a useful tool for dramatically increasing the separation power for complex separation problems, even though the operation of such a system is very demanding.[19-26] As I have already mentioned in Chapter 8, high peak capacities of about 900 can now be achieved with today's instrumentation. It should be noted that the peak capacity tells you nothing about the resolution of critical peak pairs. Although modern time-of-flight mass spectrometers have an extremely high resolving power, the structural elucidation of isomers is not possible. This can require a chromatographic separation in order to isolate these compounds. If a baseline separation is mandatory to improve the selectivity, two-dimensional liquid chromatography might play an ever more important role in the future.

Finally, I would like to conclude this chapter and monograph with some thoughts on the hyphenation of high-temperature HPLC with two-dimensional liquid chromatography. In comprehensive two-dimensional liquid chromatography (LCxLC), a second dimension column is added to the first dimension column. The principle is based on collecting fractions eluting from the first dimension column in a second loop. This fraction is then transferred to the second dimension column after the sampling period has elapsed. While this fraction is being transferred to and analyzed on the second dimension column, the eluate from the first dimension column is again collected in a second loop. This means that the analysis time on the second dimension column has to match the sampling time of the effluent from the first dimension column. Although there have been many publications dealing with LCxLC systems, the type of detector used in these studies has either been a UV or an ELS detector. However, since the samples which are analyzed in the life sciences are becoming increasingly complex, only a mass spectrometer provides the benefit of gathering structural information from the detected peaks. This means that an LCxLC system which is based on mass spectrometry for detection should be designed so that the flow rate is compatible with electrospray ionization, since this is by far the most popular ionization technique. This means that it makes no sense to employ a 4.6 mm ID column, which is operated at 5 ml min^{-1} as a second dimension column, because this is definitely too high a flow rate for MS applications. The only alternative is to use a flow splitter, but this will complicate the analytical set-up and make it more error-prone. Therefore, the flow rate of the second dimension column should be adjusted for compatibility with ESI-MS.

As was shown above, capillary liquid chromatography appears to be highly suitable for a direct coupling of the LC system with the MS detector. Hence, the column ID and thus the flow rate of the second dimension column should be adjusted accordingly. It should also be taken into account that the sampling rate of the MS should be high enough to allow for reliable quantification. As I have demonstrated, commercial TOF instruments are available with fast scanning rates, as high as 100 scans per second.

Now, high-temperature HPLC can play a pivotal role in two-dimensional liquid chromatographic separations. In a recent review article by Stoll and Carr, the benefits of high-temperature HPLC for the design of a comprehensive two-dimensional HPLC system have been described.[24] I can highly recommend this article to the interested reader to learn more about two-dimensional HPLC and the influence of temperature on the second dimension column. I will therefore not go into detail here, but summarize the most important features of high eluent temperatures in comprehensive LCxLC. By increasing the temperature, ultra-fast gradient elution separations with excellent retention time repeatability can be obtained using narrow-bore columns. If a temperature around 100 to 120 °C is considered for the second dimension separation, the viscosity of all binary solvent systems is dramatically decreased, as was shown in Chapter 4.2 (see Appendix B for a complete data overview). But this is not the only advantage. In addition, if solvent gradient programming is applied, the large viscosity maximum is also greatly reduced, which means that there is hardly any increase in the overall pressure during the solvent gradient. This means, at least theoretically, that all the solvent systems I have listed in Table 1.1 can be used for an efficient method development as well as for a screening of a phase system with the highest orthogonality. Furthermore, as was outlined in Chapter 6, columns can be used at a velocity which is much higher than the optimum velocity. Due to the flat increase of the van Deemter curve at higher eluent temperatures in the C-term dominated region, there is only a small loss in the overall efficiency when the columns are operated at very high flow rates. These are all advantages of high eluent temperatures, which can be used for a proper design of a two-dimensional liquid chromatographic system. The fear of analyte degradation can then be ignored for the vast majority of analytes to be eluted on the second-dimension column when cycle times between 20 and 30 seconds are achieved, as has been successfully demonstrated by Stoll and co-workers.[25] The authors have designed a novel 2-DL system that uses high-temperature HPLC to improve the speed of a gradient separation for the second dimension column. The authors reported a complete 25 minutes 2-DL separation of low-molecular-weight metabolites in a complex set of corn extracts, where the peak capacity of the system was about 870. The 2-D high-temperature HPLC system was based on two columns that were orthogonal in their retention mechanism, and the second dimension column could be run under high-temperature HPLC conditions to obtain a very fast cycle time. A linear gradient from 0 to 70% B was run within a cycle time of about 21 seconds, taking into account that the system re-equilibration time was only 3 seconds. Clearly, this extremely short re-equilibration time can be attributed to the fact that the column is operated at 110 °C.

What needs to be considered for further technical improvements is that the flow of the second dimension column should be adjusted to the lower $\mu l\,min^{-1}$ range. In this case, the full benefit of capillary liquid chromatography can be achieved as I have described above, which means that the high detection sensitivity of mass spectrometry can be combined with the high chromatographic resolving power of a two-dimensional separation.

I trust that I have demonstrated why high eluent temperatures, capillary LC and mass spectrometry have a huge potential for further development in liquid chromatographic separation science. I hope that in the years to come, new advances in column technology will also bolster the use of the special hyphenation techniques which I have described in Chapter 8.

References

1. H. Chen and C. Horvath, *Anal. Meth. Instr.*, 1993, **1**, 213.
2. T. Teutenberg, K. Hollebekkers, S. Wiese and A. Boergers, *J. Sep. Sci.*, 2009, **32**, 1262.

3. R. Grolik, *Effizienzsteigerung durch Hochtemperatur-HPLC in der Analytik - Ermittlung der Einsatzmöglichkeiten für die quantitative LC-MS/MS -*, Diploma thesis, Department of Chemistry, Niederrhein University of Applied Sciences, Krefeld, Germany, 2008.
4. Y. Shen, R. Zhao, S. J. Berger, G. A. Anderson, N. Rodriguez and R. D. Smith, *Anal. Chem.*, 2002, **74**, 4235.
5. Y. Saito, K. Jinno and T. Greibrokk, *J. Sep. Sci.*, 2004, **27**, 1379.
6. J. Granger, R. Plumb, J. Castro-Perez and I. D. Wilson, *Chromatographia*, 2005, **61**, 375.
7. http://www.eksigent.com/hplc/ (last accessed October 2009).
8. http://www.waters.com/waters/nav.htm?cid = 514210 (last accessed October 2009).
9. J. Bowermaster and H. McNair, *J. Chromatogr.*, 1983, **279**, 431.
10. H. McNair and J. Bowermaster, *J. High Resolut. Chromatogr. Chromatogr. Commun.*, 1987, **10**, 27.
11. K. Ryan, N. M. Djordjevic and F. Erni, *J. Liq. Chromatogr. Relat. Technol.*, 1996, **19**, 2089.
12. M. H. Chen and C. Horvath, *J. Chromatogr., A*, 1997, **788**, 51.
13. F. Houdiere, P. W. J. Fowler and N. M. Djordjevic, *Anal. Chem.*, 1997, **69**, 2589.
14. N. M. Djordjevic and F. Houdiere, *Rev. Anal. Chem.*, 1998, **17**, 207.
15. N. M. Djordjevic, F. Houdiere and P. Fowler, *Biomed. Chromatogr.*, 1998, **12**, 153.
16. T. Andersen, P. Molander, R. Trones, D. R. Hegna and T. Greibrokk, *J. Chromatogr., A*, 2001, **918**, 221.
17. T. Andersen, Q. N. Nguyen, R. Trones and T. Greibrokk, *J. Chromatogr., A*, 2003, **1018**, 7.
18. T. Andersen, I. L. Skuland, A. Holm, R. Trones and T. Greibrokk, *J. Chromatogr., A*, 2004, **1029**, 49.
19. I. Francois, K. Sandra and P. Sandra, *Anal. Chim. Acta*, 2009, **641**, 14.
20. J. N. Fairchild, K. Horvath and G. Guiochon, *J. Chromatogr., A*, 2009, **1216**, 1363.
21. J. Pol and T. Hyotylainen, *Anal. Bioanal. Chem.*, 2008, **391**, 21.
22. I. Francois, A. de Villiers, B. Tienpont, F. David and P. Sandra, *J. Chromatogr., A*, 2008, **1178**, 33.
23. P. Dugo, F. Cacciola, T. Kumm, G. Dugo and L. Mondello, *J. Chromatogr., A*, 2008, **1184**, 353.
24. D. R. Stoll, X. P. Li, X. O. Wang, P. W. Carr, S. E. G. Porter and S. C. Rutan, *J. Chromatogr., A*, 2007, **1168**, 3.
25. D. R. Stoll, J. D. Cohen and P. W. Carr, *J. Chromatogr., A*, 2006, **1122**, 123.
26. P. J. Schoenmakers, G. Vivo-Truyols and W. M. Decrop, *J. Chromatogr., A*, 2006, **1120**, 282.

APPENDIX A
Vapour Pressure Data

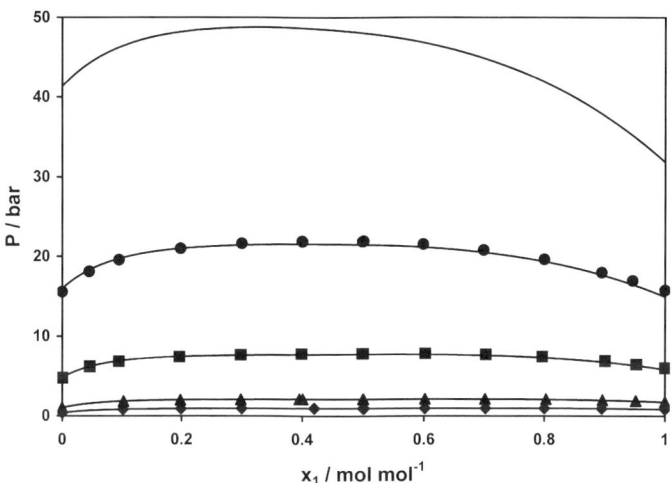

Figure A.1 Experimental isothermal P-x-data of acetonitrile (1)–water (2) at different temperatures: ♦, 75 °C; ▲, 100 °C; ■, 150 °C; ●, 200 °C; and 250 °C, correlated with NRTL (—).[1] The pressures for 250 °C have been calculated with the constants given in ref. 1. (Reproduced with kind permission from Elsevier.)

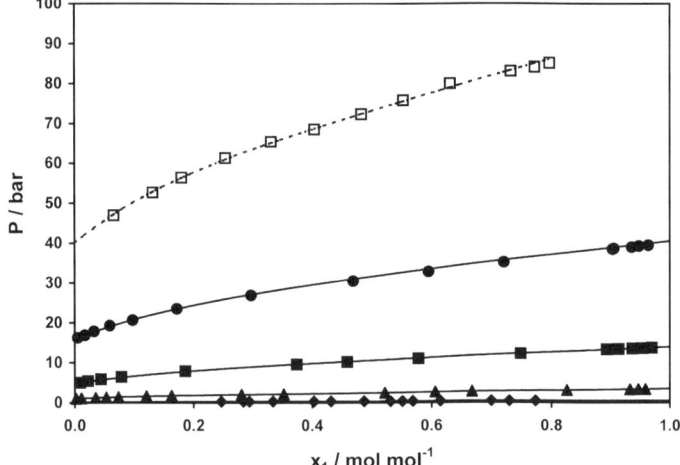

Figure A.2 Published isothermal P-x-data of methanol (1)–water (2) at different temperatures: ♦, 50 °C; ▲, 100 °C; ■, 150 °C; ●, 200 °C; and □, 250 °C, correlated with NRTL (—) or PSRK (--).[1] Literature data have been taken from DDB 2008. (Reproduced with kind permission from Elsevier.)

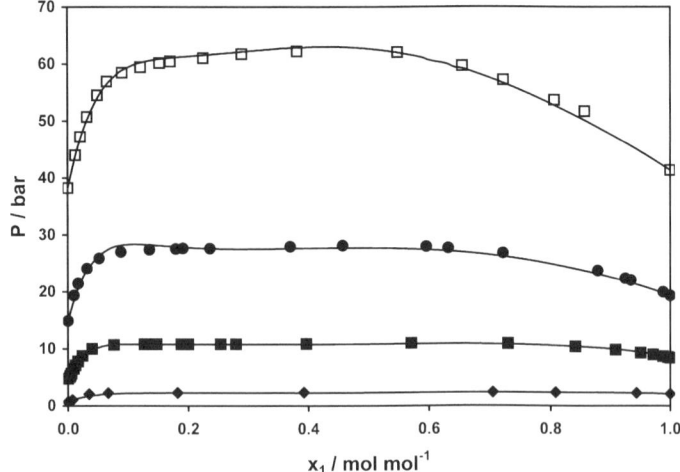

Figure A.3 Experimental isothermal P-x-data of tetrahydrofuran (1)–water (2) at different temperatures: ♦, 90 °C; ■, 150 °C; ●, 200 °C; and □, 250 °C, correlated with NRTL.[1] (Reproduced with kind permission from Elsevier.)

Appendix A

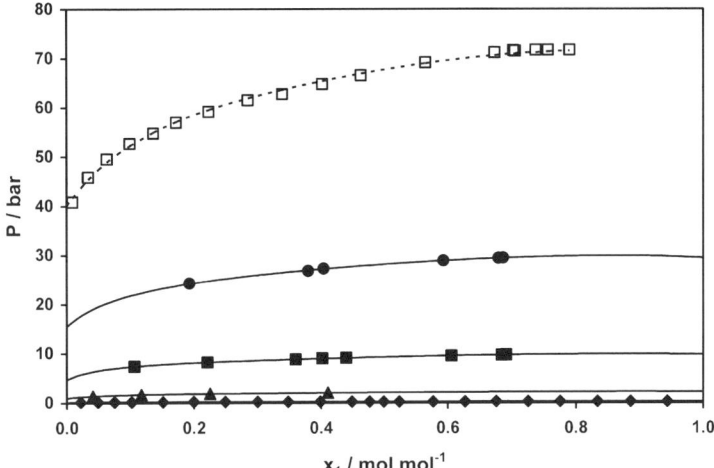

Figure A.4 Published isothermal P-x-data of ethanol (1)–water (2) at different temperatures: ♦, 50 °C; ▲, 100 °C; ■, 150 °C; ●, 200 °C; and □, 250 °C, correlated with NRTL (—) or PSRK (--).[1] Literature data have been taken from DDB 2008. (Reproduced with kind permission from Elsevier.)

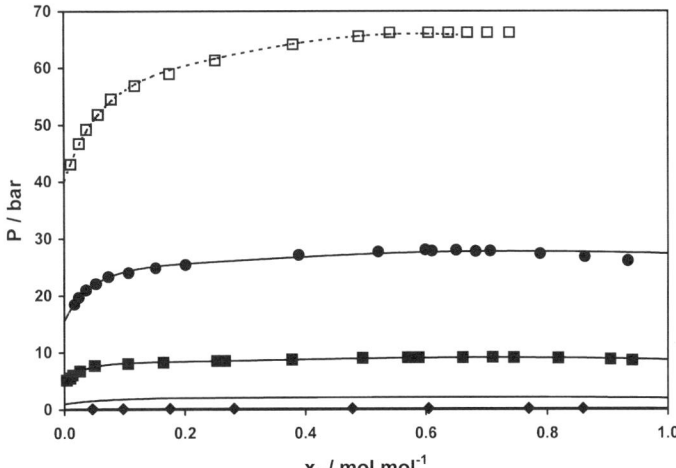

Figure A.5 Published isothermal P-x-data of isopropanol (1)–water (2) at different temperatures: ♦, 50 °C; 100 °C, only correlated; ■, 150 °C; ●, 200 °C; and □, 250 °C, correlated with NRTL (—) or PSRK (--).[1] Literature data have been taken from DDB 2008. (Reproduced with kind permission from Elsevier.)

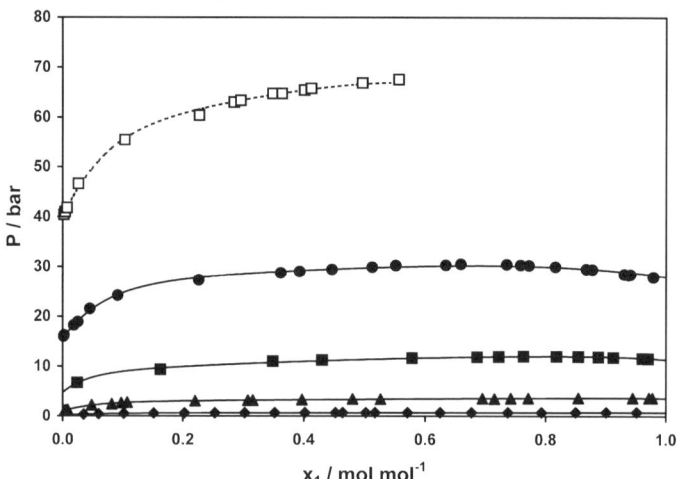

Figure A.6 Published isothermal P-x-data of acetone (1)–water (2) at different temperatures: ♦, 50 °C; ▲, 100 °C; ■, 150 °C; ●, 200 °C; and □, 250 °C, correlated with NRTL (—) or PSRK (--).[1] Literature data have been taken from DDB 2008. (Reproduced with kind permission from Elsevier.)

APPENDIX B
Viscosity Data

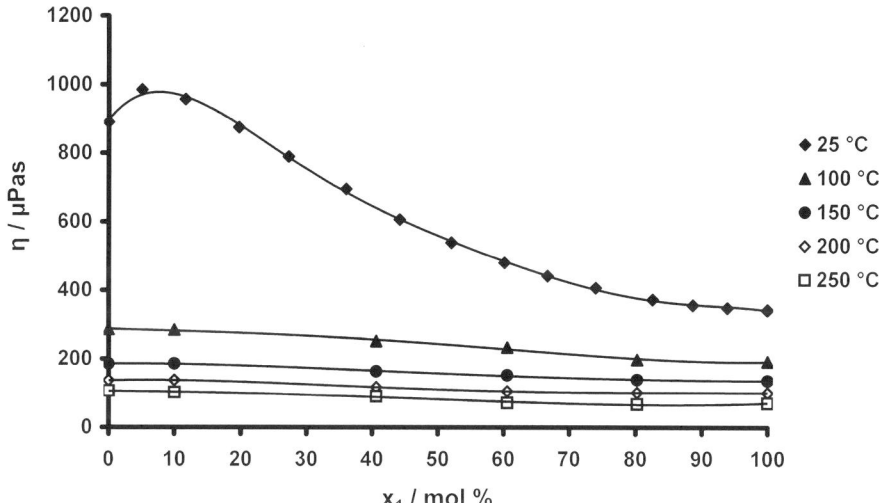

Figure B.1 Experimentally determined viscosities of the binary mixture acetonitrile (1)–water (2) at different temperatures and 100 bar.[2] Data at 2 °C have been taken from literature.

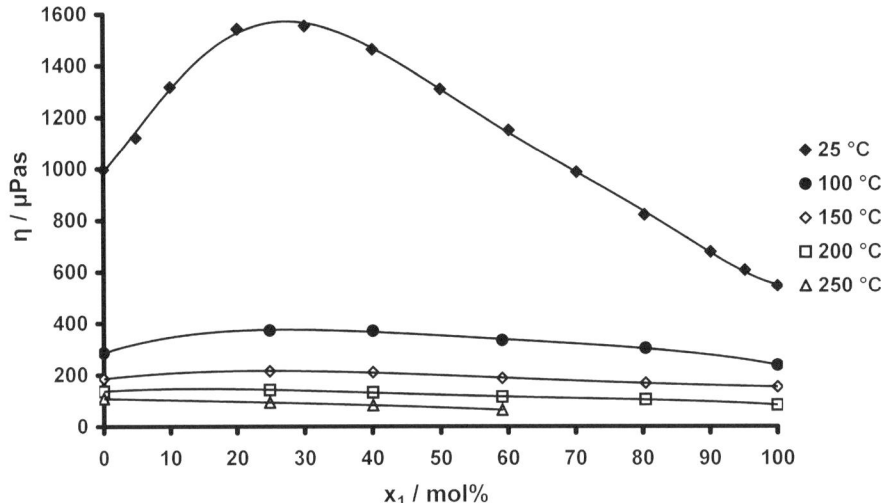

Figure B.2 Experimentally determined viscosities of the binary mixture methanol (1)–water (2) at different temperatures and 100 bar.[2] Data at 25 °C have been taken from literature.

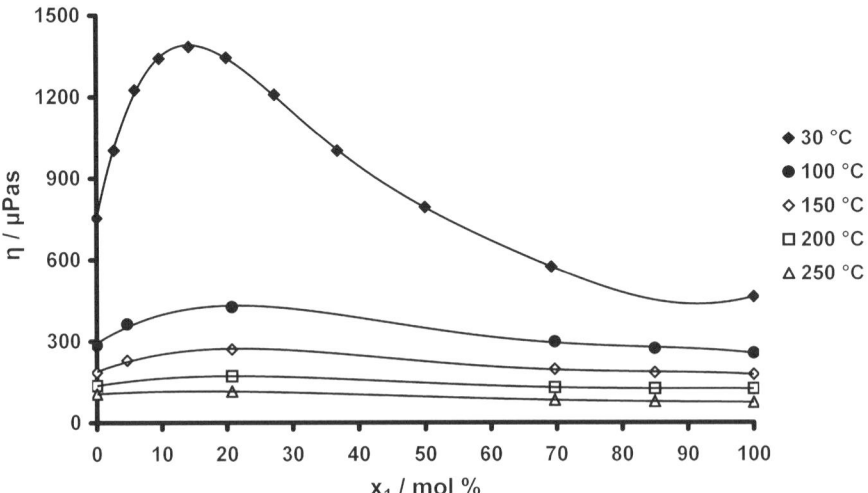

Figure B.3 Experimentally determined viscosities of the binary mixture tetrahydrofuran (1)–water (2) at different temperatures and 100 bar.[2] Data at 30 °C have been taken from literature.

Appendix B

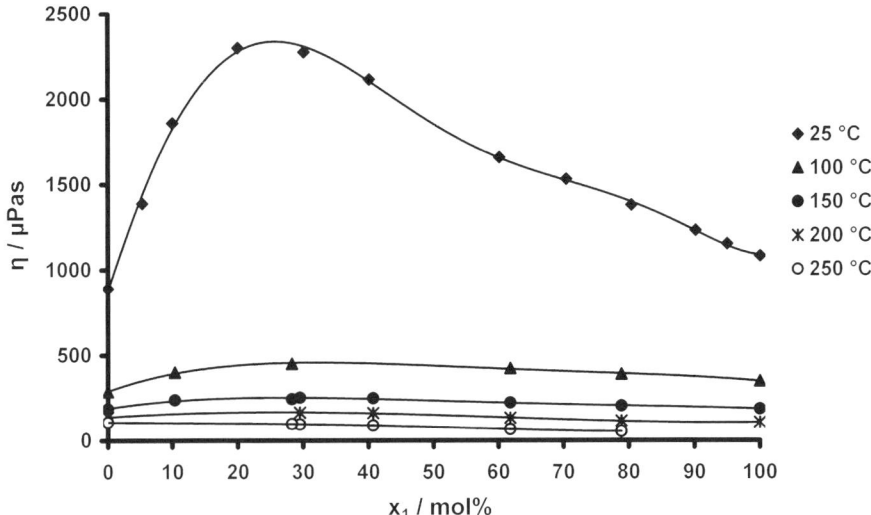

Figure B.4 Experimentally determined viscosities of the binary mixture ethanol (1)–water (2) at different temperatures and 100 bar.[2] Data at 25 °C have been taken from literature.

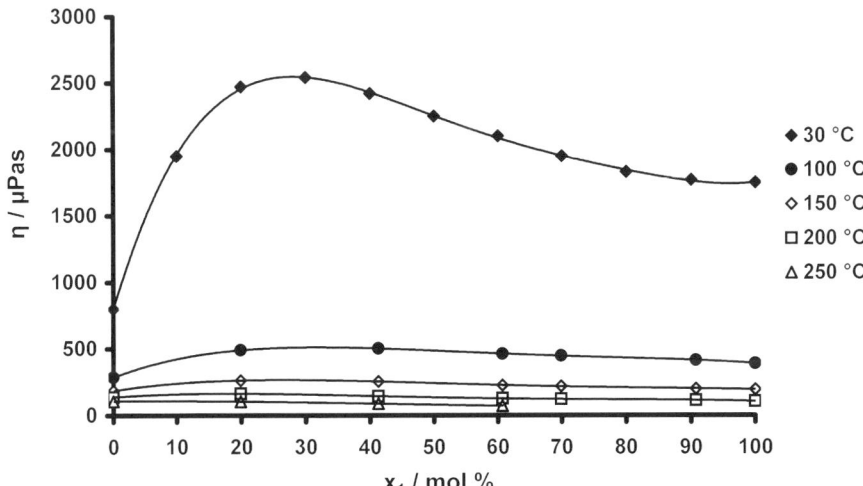

Figure B.5 Experimentally determined viscosities of the binary solvent mixture isopropanol (1)–water (2) with a pressure of 100 bar at different temperatures.[2] Data at 30 °C at atmospheric pressure have been taken from literature.

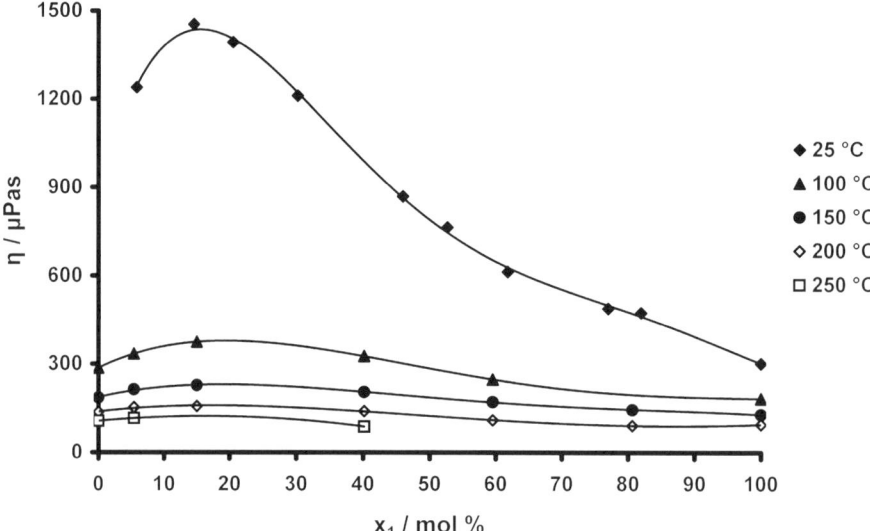

Figure B.6 Experimentally determined viscosities of the binary mixture acetone (1)–water (2) at different temperatures and 100 bar.[2] Data at 25 °C have been taken from literature.

APPENDIX C
Static Permittivity Data

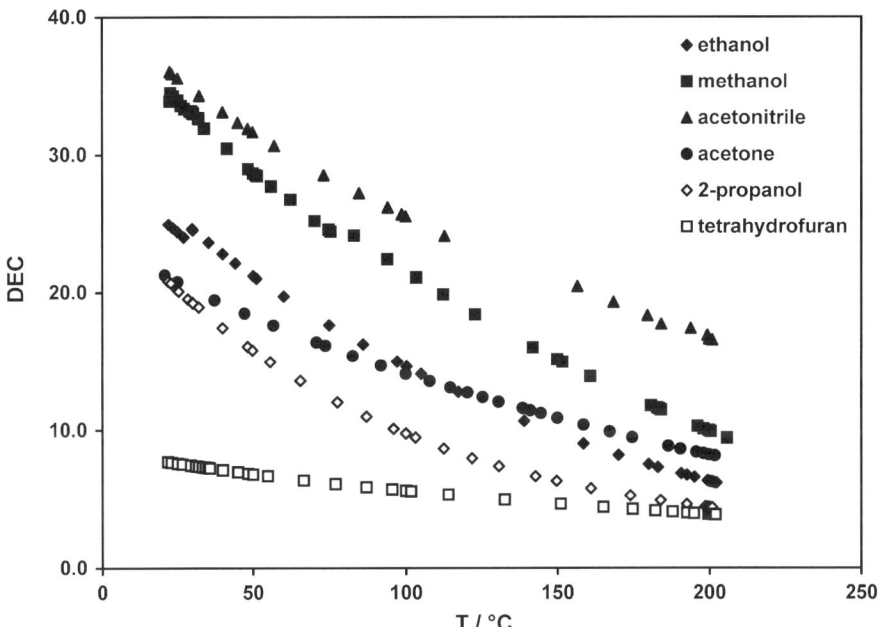

Figure C.1 Temperature dependence of the static permittivities of pure solvents at a constant pressure of 100 bar.[3] (Reproduced with kind permission from Elsevier.)

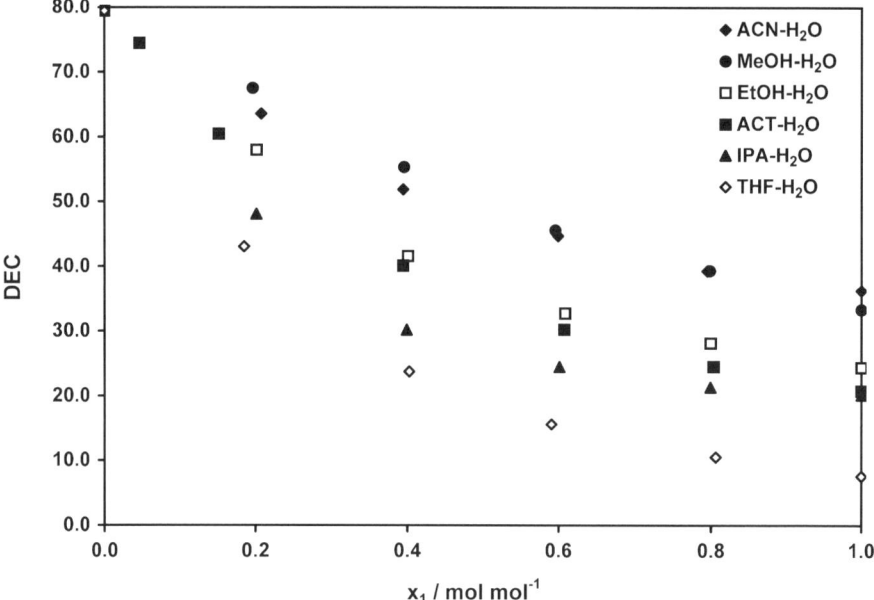

Figure C.2 Experimentally determined static permittivities of binary solvent mixtures at 25 °C and 100 bar.[3] (Reproduced with kind permission from Elsevier.)

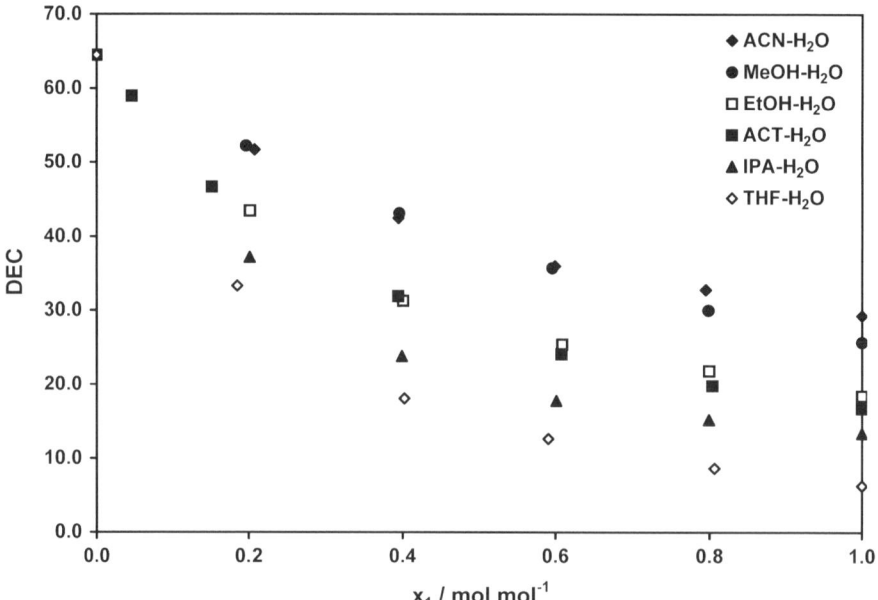

Figure C.3 Static permittivities, interpolated from the experimental data, of binary solvent mixtures at 70 °C and 100 bar.[3] (Reproduced with kind permission from Elsevier.)

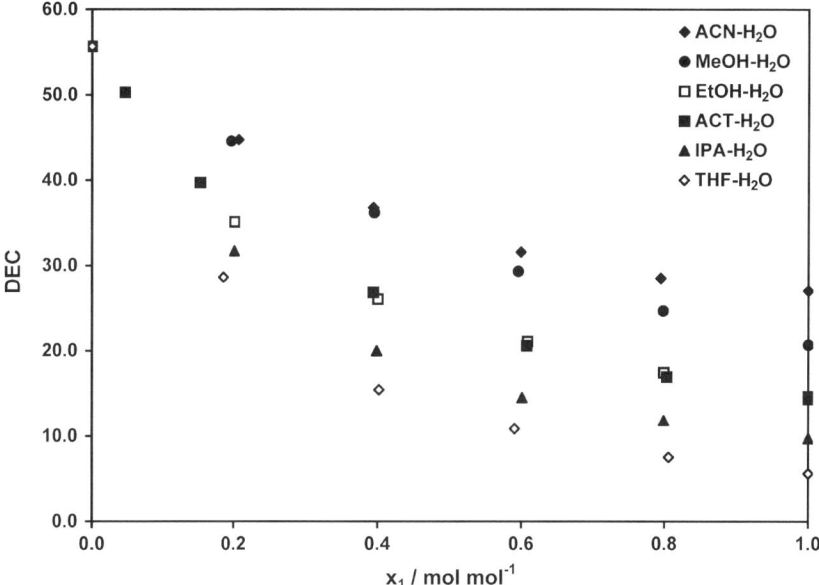

Figure C.4 Experimentally determined static permittivities of binary solvent mixtures at 100 °C and 100 bar.[3] (Reproduced with kind permission from Elsevier.)

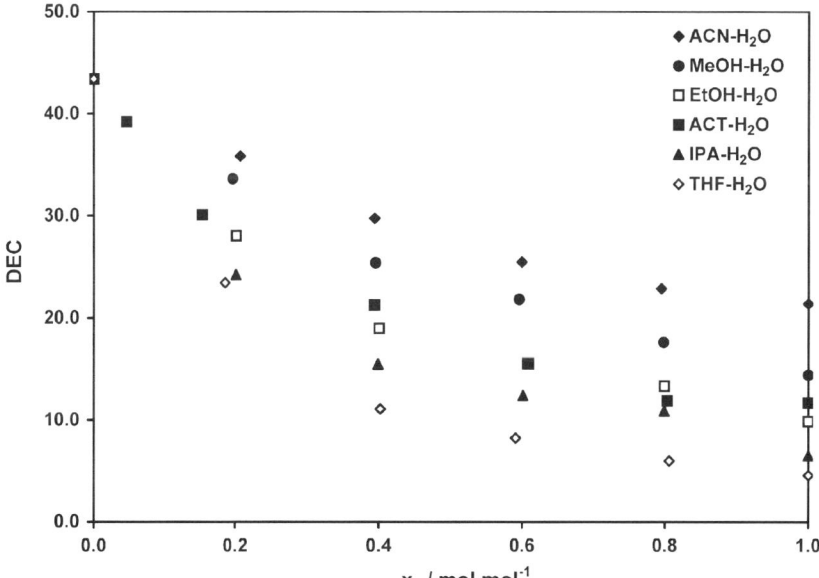

Figure C.5 Static permittivities, interpolated from the experimental data, of binary solvent mixtures at 150 °C and 100 bar.[3] (Reproduced with kind permission from Elsevier.)

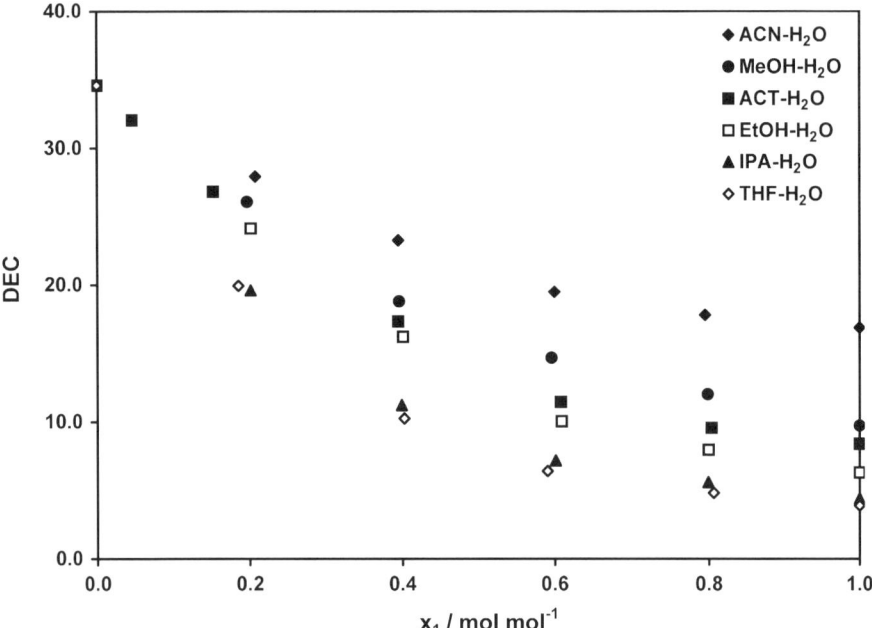

Figure C.6 Experimentally determined static permittivities of binary solvent mixtures at 200 °C and 100 bar.[3] (Reproduced with kind permission from Elsevier.)

References

1. T. Teutenberg, P. Wagner and J. Gmehling, *J. Chromatogr., A*, 2009, **1216**, 6471.
2. T. Teutenberg, S. Wiese, P. Wagner and J. Gmehling, *J. Chromatogr., A*, 2009, **1216**, 8470.
3. T. Teutenberg, S. Wiese, P. Wagner and J. Gmehling, *J. Chromatogr., A*, 2009, **1216**, 8480.

Subject Index

Page references to *figures* and *tables* are shown in *italics*.

Abbott, S. 24
acetone-water
  pressure and temperature *vs.* column dimensions 73
  pure acetone, physical properties *3*
  static permittivity *77, 201–4*
  vapour pressure isothermal P-x-data *196*
  viscosity and temperature *200*
acetonitrile-water
  pressure and temperature *vs.* column dimensions 73
  pure acetonitrile, physical properties *3*
  static permittivity *77, 201–4*
  TOF-MS 61–2, *62, 63*
  vapour pressure isothermal P-x-data *57, 58–9*, 193
  viscosity and temperature *65, 197*
alcohols
  HPLC-FID separation *162, 163*
  longer chain 72–3, 78
aldehydes and ketones, separation 78–80, *79, 80, 82*, 82–3
analyte stability
  critical criteria 154–7
  Damköhler number (Da) 149–50
  evaluation 150–2
  influence of the stationary phase 152–4
Antia and Horváth form of the van Deemter equation 128
antibiotic drugs *136, 139, 152*

back-pressure regulation, phase transition prevention 22–3, 54, 56–60
capillaries
  polyetheretherketone (PEEK) 36, 106
  stainless-steel 28, *36*, 36–7, *39*
capillary and nano HPLC 184–9, *187*
capillary-TOF-MS 184–9
carbon columns *see* pure graphitized carbon columns
carbon-clad zirconia 100–1
Carr, P. W. 20
  Carr and Thompson rules, analyte stability 154–6
Chester, T. L. 69
chromatographic techniques
  gel-permeation chromatography (GPC) 30–1, 67, 104–5
  high-pressure liquid chromatography (HPLC) 1, *55*
  high-temperature liquid chromatography (HTLC) 1–8
  reversed-phase liquid chromatography (RP-HPLC) 2–3
  size-exclusion chromatography (SEC) 104–5
  two-dimensional liquid chromatography 189–91
  unified chromatography 69
column ageing test procedure 88
column bleed 88–91

column design *see* stationary phases
column dimensions, effect on mobile phase pressure 72–4, *73*
column heating
  air-bath ovens 17, 24, 40, 46–8
  block-heating ovens 41–2, *42,* 48–9
  water-jacket ovens 24, 40–1, *41,* 48
  see also stationary phase, general aspects
column systems 20–1
cytostatic drugs *136, 139, 152*

Damköhler number (Da) 149–50
detectors 21–2, 90
  optimization 143–6
  see also specific detector system
dielectric constant *see* static permittivity
diffusion coefficient 71–2
Dortmund Data Bank (DDB) 83–5
drug screening 177–80
DryLab® software 80, 118, 120

efficiency, temperature effect on 128–31
  plate height *vs.* velocity *129*
electrospray-ionization mass spectrometry (ESI-MS) 177–80, 184–5
eluent cooling 18, 42–3, *44*
Engelhardt, H. 1
  Engelhardt test 69
equations
  Hagen–Poiseuille 67–8
  Knox and Thijssen, resolution 118
  peak area 150
  selectivity, temperature effects 125
  van Deemter equation and derivatives 6, 71, 128–9
    Antia and Horváth form 128
  van't Hoff 119–23
ethanol-water
  pressure and temperature *vs.* column dimensions *73*
  pure, physical properties *3*
  static permittivity *77, 201–4*
  supercritical conditions 117
  vapour pressure isothermal P-x-data *195*
  viscosity and temperature *199*

evaporative light-scattering detection 22, 61, 90

flame ionization detection (FID) 159–64
fluorescence detection
  and eluent temperature 18, 21–2
  vs. UV 17, 59
frictional heating *see* viscous heat dissipation
future prospects
  capillary and nano HPLC 184–9, *187*
  pellicular particles 182–3
  two-dimensional liquid chromatography 189–91

gel-permeation chromatography (GPC) 30–1, 67, 104–5
graphitized carbon columns *see* pure graphitized carbon columns

Hagen–Poiseuille equation 67–8
heating and cooling systems
  basic aspects 15–20
  block temperature *vs.* heating rate *46*
  column heating and oven systems 40–2, 46–50
  mobile phase
    column considerations 116–18
    post-column cooling 18, 42–3, *44, 47*
    preheating 17–18, *19,* 25–39
    special requirements 114–16
  temperature programming 43–6, 135–40
  see also column heating; temperature
height equivalent to a theoretical plate H(u) (HETP) 6–7, 70–1
  vs. linear velocity *7, 70*
Heinisch, S. 2
Hesse, G. 1
high-pressure liquid chromatography (HPLC)
  definition 1
  typical system *55*
high-temperature liquid chromatography (HTLC)
  definitions 1–4
  rational for 5–8
  terminologies, alternative 2

Subject Index

honey, adulterated, HPLC-IRMS separation *174*
HPLC-FID *see* flame ionization detection
hyphenation techniques 8, 42, 77, 158–9
  capillary-TOF-MS 184–9
  electrospray-ionization mass spectrometry (ESI-MS) 177–80, 184–5
  flame ionization detection (FID) 159–64
  isotope ratio mass spectrometry (IRMS) 168–74
  LC Taste® 174–7, *176*
  nuclear magnetic resonance (NMR) spectroscopy 164–8
  two-dimensional liquid chromatography 189–91

isocratic separation, temperature gradient in 135–40
isopropanol-water
  heptane–isopropanol separation *102*
  isopropanol-water system 72
  pressure and temperature *vs.* column dimensions *73*
  pure isopropanol, physical properties *3*
  static permittivity *77, 201–4*
  vapour pressure isothermal P-x-data *195*
  viscosity and temperature *199*
isothermal and isocratic separation, steroids 131–5, *132–4*
isotope ratio mass spectrometry (IRMS) 168–74

ketones and aldehydes, separation 78–80, *79, 80, 82,* 82–3
Knox and Thijssen resolution equation 118

LC Taste® 174–7, *176*

metal oxide columns *see* zirconium dioxide-based stationary phases
methanol-water
  high temperature column purging 92
  pressure and temperature *vs.* column dimensions *73*
  pure methanol, physical properties *3*
  static permittivity *77, 201–4*
  vapour pressure isothermal P-x-data *58, 194*
  viscosity and temperature *66, 198*
mobile phase
  post-column cooling 42–3
  preheating 17–18, *19,* 25–39
  system pressure, temperature and column dimensions 52–64
  temperature and vapour pressure 52–64
Molnár Institute for Applied Chromatography 123

nano HPLC 184–9, *187*
Neue test 69, 92
non random two liquids (NRTL) model 84
nuclear magnetic resonance (NMR) spectroscopy 164–8

on-flow NMR 165
ovens systems *see* column heating

peak area
  analyte stability evaluation 150–2
  equation 150
  vs. temperature *151, 152*
  see also van't Hoff equation
peak splitting 33
pellicular particles 182–3
Peltier element, cooling 42–3
pesticides, capillary HPLC separation 187–9, *188, 189*
pharmaceutical compounds, separation 68
phase transition prevention
  back-pressure regulation 22–3, 54, 56–60
  using a restriction capillary 61–4
phenols, HPLC-FID separation *162*
plate height (H)
  plate height *vs.* velocity *129*
  see height equivalent to a theoretical plate $H(u)$ (HETP)

PLRP-S column
   on-column degradation of thalidomide *155*
   retention factors *122*
polar retention effect on graphite (PREG) 106–7
polycyclic aromatic hydrocarbon (PAH)
   effect of eluent preheating 17–18, *19*
   peak area *vs.* eluent temperature 42–3, *44, 144*
   retention times *121*
   separation conditions *16*
polyetheretherketone (PEEK) capillaries 36, 106
polymeric stationary phases 104–5
predictive Soave-Robinson-Kwong (PSRK) model 84
preheating, mobile phase
   effects 17–18, *19*
   efficiency assessment 37–9
   eluent temperature measurement *38–9*
   technical considerations 35–7
   thermal mismatch broadening 25–31
   viscous heat dissipation *26,* 31–5
   see also heating systems; temperature
propanol *see* isopropanol-water
pure graphitized carbon columns 105–7
   polar retention effect on graphite (PREG) 106–7

resolution
   Knox and Thijssen equation 118
   temperature effect on 15–17, 118–19
restriction capillary
   kinetics and column pressure, practical implications 69–74
   for phase transition prevention 61–4
   viscosity, practical implications 67–9
retention
   enthalpy 119–25
   experimental *vs.* simulated retention times *121, 124*
   linear/non-linear responses *120*
   retention factor *122*
   temperature effects 119–25
   temperature-gradient separation *124, 137, 139*

reversed-phase liquid chromatography (RP-HPLC), definition 2–3

selectivity, temperature effects 125–7
   equation 125
   van't Hoff plots *127*
silica-based stationary phases 93–7
   characteristics 87
   pellicular particles 182–3
   reversed-phase 5, 21
   temperature stability 5, 20–1, *94–6*
   temperature/pH interaction 93–7
size-exclusion chromatography (SEC) 104–5
Smith, Roger M. 2, 8
solute retention 33–4, *34*
solvents, pure
   physical properties *3*
   static permittivity *76*
solvents, static permittivity of mixtures *77*
'sputtering' 62
stainless-steel capillaries 28, *36,* 36–7, *39*
static permittivity (DEC) 8, *201–4*
   influence of temperature 8, *75–7,* 75–80, 84, 158, *201–4*
stationary phase, general aspects
   batch-to-batch reproducibility 87
   column ageing test procedure 88
   column bleed 88–91
   columns and temperature maxima *110*
   design difficulties 88–9
   detector response in different columns *90–1*
   high temperature degradation 91–3
   influence on analyte stability 152–4
   see also column heating; specific stationary phase components
steroids
   isothermal and isocratic separation 131–5, *132–4*
   peak areas *151*
   thermal mismatch broadening 25–6
stop-flow NMR 165
sulfonamides
   isothermal and isocratic separation 140–3, *141–2*

Subject Index

retention times *120*, 123
temperature-gradient separation *124*, 126
system
  basic set-up 15–23
  capillary and nano HPLC 186–9, *187*
  continuous-flow for ESI-MS 177–80, *178*
  general requirements 8–9
  HPLC-FID 160–1
  HPLC-IRMS *169*
  HPLC-NMR 165–6, *167*
  LC Taste® 174–7, *176*

temperature
  columns and temperature maxima *110*
  effect on analyte stability 149–57
  effect on efficiency 128–31
  effect on resolution 15–17, 118–19
  effect on retention 119–25
  effect on selectivity 125–7
  high temperature column degradation 91–3
  mobile phase temperature and vapour pressure 52–64
  range 4–5
  temperature programming 43–6, 135–40
  see also heating and cooling systems
temperature-gradient separation
  simultaneous temperature and solvent gradient separation 140, 143
  sulfonamides and uracil *124*
  using HPLC-IRMS 171–2, *173*
testosterone and related hormones
  HPLC-IRMS separation 172, *173*
  retention and peak shape *27*
  thermo-responsive separation *108*
tetrahydrofuran (THF)-water
  miscibility gap 116–17
  pressure and temperature *vs.* column dimensions *73*
  pure tetrahydrofuran, physical properties *3*
  solvent system *81*, 81–3
  static permittivity *77, 201–4*
  vapour pressure isothermal P-x-data *60, 194*
  viscosity and temperature *198*

thermal mismatch 25
  broadening 25–31
  column thermal effects *26*
  contact heating 28–9
  preheating coil length and flow rate 26–31, *27, 28, 29*
thermo-responsive stationary phases 107–9, *108–9*
Thompson, J. D., Carr and Thompson rules, analyte stability 154–6
time-of-flight mass spectrometer (TOF-MS) 61–2, *62, 63, 68*
titanium dioxide stationary phases
  temperature and chemical stability 101–4, *102–3*
  see also zirconium dioxide-based stationary phases
trimethoprim, non-linear retention *120*
two-dimensional liquid chromatography 189–91

unified chromatography 69
UV *vs.* fluorescence detection 17, 59

van Deemter equation and derivatives 6, 71, 128–9
van't Hoff equation 119–23
vapour pressure
  data 193–6
  liquid water *53*
  and temperature 52–64
viscosity
  data 197–200
  influence of temperature 7–8, 64–74
  maximum 7–8
viscous heat dissipation
  definition 31–2
  effect on flow profile *26*, 32–5

water
  binary mixes
    pressure and temperature *vs.* column dimensions *73*
    static permittivity *77, 201–4*
    tetrahydrofuran (THF)-water miscibility gap 116–17

vapour pressure isothermal P-x-data
193–6
viscosity and temperature
197–200
pure
physical properties *3*
static permittivity and temperature *75*
vapour pressure *53*
viscosity and temperature *64*
water only separation 21

zirconium dioxide-based stationary phases
as alternative to silicates 20
and analyte stability 150–2
elution of sulfathiazole on *153*
on-column degradation of thalidomide *154*
polymer-coated 98–101
retention mechanisms 87–8
steroids, isothermal and isocratic separation 131–5
temperature stability 97–101, *99*